Praise for Richard N

Wild Waters: The Magic of Ireland's Rivers and Lakes
'A vital exploration of our beautiful inland waterways.' Fergal Keane

'A timely and deeply felt contribution to Irish natural history.' Michael Viney

'A beguiling cocktail of personal experience, historical fact and the wonder of nature.' *RTÉ Guide*

'The latest in an extraordinary contribution to environmental literature in this country.' *Sunday Independent*

'Always a joy to read as he explores Ireland's natural beauty.' *Westmeath Examiner*

Wild Shores: The Magic of Ireland's Coastline
'Takes us into places we would never have the time (or sea legs) to reach.' Catherine Cleary

'Well-researched and beautifully written. An absolute gem.' Zoe Devlin

'An affectionate and timely celebration of Ireland's richly varied coastline.' Bryan Dobson

'A great read – whatever part of the coast you visit.' Éanna ní Lamhna

'His description moves me. An engaging and ambitious new book.' Michael Viney

'An exquisite, thought-provoking odyssey.' *Westmeath Chronicle*

Wild Woods: The Magic of Ireland's Native Woodlands
'A book to inspire anyone.' Michael Viney

'A story of dogged persistence and patience rewarded.' *RTÉ Culture*

'Read and be enchanted and informed.' *Small Woods Magazine*

'An exhilarating read.' *Woodland Magazine*

Dublin Bay: Nature and History

'A very keepable book.' *The Irish Examiner*
'A rich story, full of colour.' *The Irish Independent*
'Has encyclopaedic authority and the comforting accessibility of a classic.' *Sunday Times*

Bird Habitats in Ireland

'A truly remarkable book, highly readable.' *British Trust for Ornithology*
'An excellent volume that I can recommend strongly to all birdwatchers and ecologists in Ireland and elsewhere.' *Proceedings of the Royal Irish Academy*
'Will be an invaluable tool for future ornithologists and an interesting read for anyone with a concern for birds.' *Biodiversity Ireland*

Ireland's Coastline: Exploring its Nature and Heritage

'Should be in every library and school and, if possible, in every home.' *The Sunday Business Post*
'As much a photojournalistic poem as a glossily informative hardback.' *The Irish Independent*
'A marvellous guide – a definitive volume.' *The Irish Examiner*
'Succeeds in rekindling the wonderful fascination of the seashore.' *Irish Naturalists' Journal*
'A major celebration of our island's diverse shores.' *Irish Mountain Log*
'Breathtaking photographs and informative prose.' *The Midwest Book Review, Oregon*

Wild Wicklow: Nature in the Garden of Ireland

'Makes the nature of the garden county accessible and interesting for every reader.' *The Irish Times*
'Sets a high standard for a county natural history.' *The Cork Examiner*

Also by Richard Nairn

Wild Wicklow: Nature in the Garden of Ireland
Ireland's Coastline: Exploring its Nature and Heritage
Bird Habitats in Ireland (joint editor)
Dublin Bay: Nature and History
Wild Woods: The Magic of Ireland's Native Woodlands
Wild Shores: The Magic of Ireland's Coastline
Wild Waters: The Magic of Ireland's Rivers and Lakes

FUTURE WILD
First published in 2024 by
New Island Books
Glenshesk House
10 Richview Office Park
Clonskeagh
Dublin D14 V8C4
Republic of Ireland
www.newisland.ie

Print ISBN: 978-1-83594-001-3
eBook ISBN: 978-1-83594-002-0

British Library Cataloguing in Publication Data. A CIP catalogue record for this book is available from the British Library.

Set in 11.5 on 16.8pt Sabon

Typeset by JVR Creative India
Edited by Sheila Armstrong
Indexed by Eileen O'Neill
Cover design by Anna Morrison, annamorrison.com
Printed by Opolgraf Printing House, Poland, opolgraf.com.pl

New Island Books is a member of Publishing Ireland.

10 9 8 7 6 5 4 3 2 1

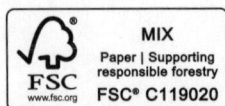

FUTURE
WILD

For my grandchildren, Ivy, Iris, Robyn, Phoebe and Lúcio.
To help restore your country for the future.

FUTURE
WILD

Nature Restoration
in Ireland

RICHARD NAIRN

NEW ISLAND

Contents

Introduction

I am standing at the front door of the house where I live. It is a bright day in May. The sun creeps up above the ridge to the east as it rises from the sea. Immediately below me is a meadow that sweeps away down the hill, bright with wildflowers in the morning sun. At the bottom of a long hill lies the old woodland where I can hear a woodpecker drumming on a dead branch. Beside it is a burgeoning plantation of native trees that is beginning to merge with the old woodland. I can hear the rushing river in the wood and imagine its winding path through the ancient trees. In the distance is a craggy hill covered in heather and flanked by dark green forest. The sun is lighting up one side of the hill while the other remains in shadow. Over me drifts the first red kite of the morning, effortlessly soaring above the meadow as it searches for its first meal of the day. I am surrounded by nature and beauty.

A decade ago, I was presented with the opportunity to buy this small farm in the valley where I live in the foothills of the Wicklow Mountains. I had always wanted to plant trees, to surround myself with nature and to be able to observe the natural world changing from day to day. Within a year I suddenly found myself with the keys to the gate. It was pure exhilaration to wander at will, to observe the land in all its diversity. The pasture had been grazed by horses and sheep and was shorn to the length of a mown lawn. The patch of old woodland at the foot of the hill was a mysterious, magical place, but the sheep were wintering there and browsing the undergrowth. Deer were also browsing in the woodland and emerged occasionally to graze in the pasture.

Spring is my favourite time of year. All is new in nature after the long cold months of winter when the struggle to survive takes precedence over everything else. As I stand here in the brightening dawn, I celebrate the return of heat from the sun. Blackbirds are singing loudly from the trees all around and wrens belt out their incessant calls from the undergrowth. The meadow is filled with butterflies and bees all working to ensure the next generations will replace them in a short time. They feed on the abundant wildflowers that have returned here since the sheep and horses were removed. Just a short walk from my front door, badgers are sleeping underground, their growing cubs venturing out at sunset to explore the woodland and meadow that their ancestors have known for hundreds of years. A solitary deer stands in the meadow, its ears twitching to listen for danger, ready to race to the cover of the trees. Nature is alive here.

Although it is not a wilderness, but a landscape managed by humans for thousands of years, this is the place that I love. It is like a painting, a panorama that stretches from east to west. I never tire of gazing out to watch the clouds that drift across, sometimes bordered in gold, sometimes carrying the threat of rain that speeds in from the west. At night the moon follows the setting sun picking out the hillside above with a ghostly light. Then I listen to the calls of a family of owls and watch them dodging in and out of our growing plantations. I want to intimately know every wild creature and wild plant here. My aim from the start was to restore nature on this land and make it a refuge for wildlife in a rapidly changing world. The red kite, extinct in Ireland for 300 years, yet now gently drifting across the fields, is a sign that nature restoration can work. I want to give back to nature, as much as possible, the beauty and diversity that has been modified and simplified by landscape change. To live with nature instead of trying to exploit it.

Nature for me is both an inspiration and a refuge. It gives me endless pleasure to immerse myself in the wild places and find there the animals and plants that still live the way they have done for thousands of years. In hard times, such as the recent pandemic, nature provides a safe anchor, its plants and animals changing with the seasons just as they have always done. All through my life, I have worked to protect and restore nature. Half a century ago, I learnt the practical methods when I worked as a nature reserve warden. As a staff member of an environmental organisation, I focussed on birds which are such a beautiful and accessible part of our wildlife. Later, working with other environmental specialists, I strived to prevent damage to nature due to the rapid development of our country. Now, in my retirement, I spend most of my time managing my own small farm with nature restoration as the main objective.

Inspiration and peace are not the only gifts that this little patch of land offers me. It provides the wood for the fire that heats my house in the winter. I spend long hours in my woodshed splitting and stacking logs to season for next winter's fuel. The sun powers the solar panels on the roof that provide most of the electricity that I need. Rain falling on the same roof drains into large tanks, a free rainwater harvest for the gardens in a dry summer. The soil in the gardens produces a variety of tasty and nutritious vegetables and fruit that will last for the whole year. Season after season, these gifts are generously given by the land that asks nothing in return except care and protection to go on producing. The American writer Aldo Leopold wrote, 'When we see land as a community to which we belong, we may begin to use it with love and respect.'[1]

The land surrounding my farm is not a natural landscape. From my front door, I can see intensive farmland and commercial forestry. I can hear the noise of machinery at a nearby quarry. Repeated ploughing and fertiliser applications have replaced the natural grassland in the valley with cereals and silage fields and confined the wildflowers to little corners that cannot be grazed or cultivated. Chemicals run off the fields into the river, threatening the aquatic life that depends on clean water. The ancient woodlands have largely been replaced by fast-growing ranks of conifers that are repeatedly clear-felled and replanted. Here there are uncontrolled herds of non-native deer, browsing the undergrowth and preventing the natural vegetation from returning.

Since acquiring this farm, I have set about managing it to restore the greatest natural diversity possible in an otherwise intensively managed landscape of farmland and forestry. 'But why would anyone abandon land to nature on a good productive farm, with fertile, well-drained soils? Why not grow a crop or graze some livestock here? Why leave the meadow alone to flower and set seed when the hay is not used when it is mowed? Why not harvest the old timber in the woodland and plant it with commercial conifers?' These are questions often posed by critics or doubters. I have made a personal decision to give priority to nature, one that is increasingly necessary in light of the biodiversity emergency that we face right now. I am attempting to restore nature on this small patch of land or, rather, to help it to restore itself. If the heavy pressures of modern land use are removed, natural succession will start to kick in.

Every gardener knows that if you dig a patch of garden and then go on your summer holidays, the ground will be covered in weeds when you return. Thistles, nettles, docks and many other smaller wild plants will reclaim this space. These native wildflowers are simply plants in the wrong place, opportunists that fill the space left by any disturbance of the ground. Richard Mabey described them as 'part of nature's immune system, of its instinctive drive to green over the barrenness of broken soil'.[2] If you abandoned your home altogether, the disturbed patch would quickly sprout with brambles, buddleia or gorse. If nobody moved into the property, within a year or two, scrub would be filling the garden, closing off paths, scrambling over fences and blocking windows. Small mammals would move in, hedgehogs might shelter in denser patches and foxes might make a den beneath the old garden shed. After a decade of 'neglect', young tree saplings would be pushing through the thorny scrub, reaching for the sunlight. The tree seeds would have fallen in the autumn, blown from mature trees in surrounding gardens or hedges, germinating and pushing up through the soil in spring. If the garden was completely abandoned to nature, it would become woodland after 20 or 30 years. This is a form of natural succession that happens before your eyes. But we rarely want this to overwhelm our beloved gardens. The continual disturbance involved in gardening interrupts the process of natural succession and moves it back to an earlier stage. It reduces the diversity of wild plants and animals and ensures that only those species that we favour can establish.

If cultivation or grazing is abandoned on farmland, as sometimes happens when the land is unproductive or the farmer gets a job elsewhere, brambles push out from the surrounding hedges and, within a year or two, shrubs and trees begin to establish in the fields. Again, natural succession is moving the fields back to woodland, mimicking a time before humans began to clear the forests and farm the land. Similarly, if grazing pressure by sheep and deer on mountain land is reduced, first heather and then trees will start to return and the succession to woodland begins again. Many other modern land uses – drainage of rivers and lakes, exploitation of peatlands, quarrying hillsides, turning sand dunes into golf courses, dumping on coastal saltmarshes or building roads and urban developments – interrupt natural succession, removing wild plants and animals and impoverishing biodiversity. All of these land uses have led to a major loss and disruption of biodiversity over many centuries. Sensitive plants and animals have become rare or even disappeared altogether from the country. Habitats like forest plantations have become uniform and species-poor, removing the natural patchworks and favouring the types of landscape that serve our burgeoning population and its needs. If the pressures can be removed, the process of natural succession can begin again.

In my lifetime, nature in Ireland has undergone a catastrophic decline. The flower-rich hay meadows of my childhood are almost a thing of the past. The ancient woodlands have been whittled away and there are only a few tiny fragments left. Most of the midland bogs have gone, their priceless peat used for gardens or gone up in smoke. Only a handful of rivers remain in a pristine condition, while the salmon and eels which used to return from the ocean each year have been decimated. Overfishing has scoured the seas around our coast, and this has already had serious implications for seabirds and marine mammals as well as the fishing communities. A quarter to a third of all species that have been assessed by the National Biodiversity Data Centre are threatened with extinction in Ireland. No longer can I see the great variety of wild plants and animals that my parents and grandparents took for granted. Crumbling beneath the pressures of modern land uses and mismanagement of the seas, wild habitats are disappearing before our eyes, and some common species are becoming rarities or slipping away altogether. Nature conservationists have been struggling for half a century to stem some of these losses. But are they simply cataloguing the destruction or, worse still, moving the deckchairs around while the ship sinks slowly beneath the waves? It is time for a fresh approach to rescuing nature.

Until very recently the philosophy and practice of nature conservation in this country have been about hanging on to the patches of semi-natural habitat and some rare species that have been inherited from the past. In this sense, nature conservation has much in common with the preservation of historic buildings, protection of archaeological sites or rare works of art. This is why natural and cultural treasures are often referred to together as 'heritage'. However, natural ecosystems are not historic relics but living and constantly changing entities. Protecting them is not about preserving specimens from the past but about helping them to evolve and change in the future. Conservation must look forwards rather than backwards.

Restoration of land for nature requires both a change in direction of land use and ongoing management to ensure that the negative human impacts on wildlife and habitats are suppressed. The more recent term 'rewilding', on the other hand, is a radical concept that also aims to remove the pressures of intensive land uses. It allows natural succession to take an unpredictable course with no clear objective except to allow change without our intervention. In a few cases, it involves reintroducing extinct or depleted species – predators such as eagles or wolves – and even old breeds of cattle and horses as functional equivalents for ancient herbivores that are now extinct. This is an attempt to replicate the complicated food webs that would have been present before their removal. Restoration and rewilding are often confused. Paddy Woodworth has written about the critical distinction, saying, 'It's common now to find the two terms used interchangeably, but they come from different contexts and carry very different implications.'[3] The fundamental difference is that restoration requires continual management from people to ensure that nature can survive and prosper.

Internationally, rewilding has developed a popular following and, while this is welcome, it is not the best approach in some parts of the world. In vast areas like the oceans or the polar regions, where active management is not practical, it is meaningful to withdraw all exploitation and allow nature to take the lead. In a country like Ireland, where little remains of the natural ecosystems and where intensive land use for thousands of years has divided the country into small parcels with multiple ownerships, the process of rewilding would be difficult or impossible to achieve on a large scale. Farmers, foresters, turf-cutters and fishers have to make a living and they are slow to change what they regard as 'traditional' uses. They often see nature as something to be exploited or, at best, ignored. There is a fundamental resistance to 'abandoning' land that has been hard-won by farming or forestry. But restoration does not mean abandonment.

Nature restoration in certain habitat types requires intervention to restart the natural processes and restore ecological functions that have been interrupted. I organised the planting of over 7,000 native trees on our farm and, after a little initial management, I can now stand back and watch them growing to form a mature woodland. But the fences around them must remain, at least for the present, while there is a large population of invasive deer in the area, as these would quickly reverse any progress in woodland establishment. After the livestock were removed from our fields, the meadow blossomed with dozens of wildflower species that had previously been suppressed and thousands of butterflies and bees appeared to pollinate the flowers. But the meadow still has to be managed, the grass cut at the end of summer and the hay removed to ensure that the more vigorous grasses do not take over. On a larger scale, rewetting peatlands that have been damaged by drainage and removal of surface layers of turf can accelerate the process of moss growth that ultimately leads to peat formation. But the rewetted bogs need careful management, adjusting water levels to ensure that they are just right for the plants to grow. Actively growing peat bogs absorb rainwater, reducing siltation in rivers and preventing flooding downstream. This is nature restoration in practice.

Removing the barriers, such as weirs and dams, in rivers allows migratory fish to spawn in the upper reaches again. Restoring river meanders slows the flow and encourages the natural sequence of riffle, glide and pool favoured by many invertebrates and fish. Allowing woodland to establish on riverbanks prevents erosion and silting of the riverbed and may ultimately slow the flow as woody debris falls into the channel. It also provides corridors for secretive animals like otters to move through the landscape. Restoring nature on the coast also involves the removal of barriers such as rock armoury, piers and sea walls to reestablish the dynamic exchange of sand between beaches and sand dunes and the growth of saltmarsh in sheltered parts of estuaries. Ending the heavy trawling in some parts of the sea around Ireland would allow fish populations and the marine ecosystem to recover with ultimate benefits for fisheries. Restored peatlands, woodlands, saltmarshes, seagrasses and native oyster beds are now seen as nature-based solutions to climate change and as hotspots for biodiversity. The goals of nature restoration need to be subdivided into tailored solutions for different habitats and species. Only in true wilderness can nature now achieve its true potential without our intervention.

Today we know that human impacts have reached into every part of this small planet and in some places have modified our environment so fundamentally that it seems to be losing the ability to support us and all the other species that live here. Up to the mid-20th century, it was largely believed that there were no limits to our use of the Earth. After World War II, the modern age of chemicals, oils, plastics and mechanisation took over and fuelled this increasing exploitation. There is no doubt that human impacts on the landscape and wildlife of Ireland have modified it almost beyond recognition. Natural forests have been virtually eliminated, the natural fertility of soils has been dramatically reduced by intensive farming and fertiliser runoff has reduced river water quality. Most of the large native grazing animals have been replaced by livestock with many of the larger predators also exterminated by centuries of persecution. Rivers have been highly modified by dams, weirs and culverts and many coasts have been so altered by artificial constructions that the natural processes of erosion and accretion no longer work effectively. In some places urban sprawl has replaced the natural habitats which disappeared long ago.

Among the principal reasons for the loss of biodiversity is a group of species introduced over the millennia from all around the world to places where they never occurred before. Some are relatively benign, fitting in with the local environment with few problems. The red squirrel is known to have been introduced to Ireland by the

Normans around the 12th century, but possibly became extinct due to the decimation of Irish woods and had to be supplemented from the English population in the 18th century. It seems to have adapted to new conifer plantations without any detrimental impacts although it has suffered due to competition from the more recently introduced American grey squirrel.[4] The worst of these new arrivals are the invasive species, uninvited guests which spread uncontrollably once released into the new environment, threatening native species or habitats with destruction. Even in some protected areas such as Killarney National Park, the ancient oak and yew woodlands are infested with rhododendron, planted originally in the nineteenth century. If allowed to spread here it threatens the future of these important ecosystems and all the species that they support, largely by shading out any regeneration of native vegetation. If rewilding involves simply avoiding any intervention, then the rhododendron spreads everywhere, shading out and competing with the seedlings of native plants and leaving only ageing and decaying trees. If we simply allow invasive deer populations to multiply without control, then we may never see the return of native woodland to the hills and valleys. So, the control of invasive species needs to be given a high priority in nature restoration projects to avoid further loss of biodiversity.

One of the questions often asked about nature restoration is: what historic stage in the development of a constantly changing landscape is restoration attempting to replicate? It is the 're' in restoration that suggests going backwards but, in truth, we can only go forwards, while being aware of what we have lost from the past. Unlike my children and grandchildren, I am old enough to remember a time before television, before computers and the internet, before mobile phones. In those days we communicated by landline or post. If we were on holiday, we sent a message home on a postcard, not by social media posts. Today, most people consider instant communication a normal part of daily life. We have almost forgotten what it was like in 'the old days'. This phenomenon is called 'generational amnesia'. In his essay 'The Unraveling of America', the author Wade Davis wrote, 'fluidity of memory and capacity to forget is perhaps the most haunting trait of our species'. In the context of our memories of natural features, we call this 'shifting baseline syndrome'. What we consider to be a healthy environment now, past generations would have thought degraded. What we judge to be degraded now, the next generation will consider to be healthy or 'normal'.

Birds offer a useful example of shifting baseline syndrome. They are among the best and most consistently recorded animals as ornithology became a popular pursuit among members of the wealthy and idle classes as long ago as the nineteenth century. In his landmark book *The Natural History of Ireland* (1850), William Thompson detailed the status of the corncrake or land-rail as it was known in his time. He wrote, 'Everywhere that we go in this island in the months of May, June and early July, except to the mountain tops or to the stony and heath-covered tracts, the call of the corncrake is heard, not only at its favourite times, in the evening and during the night, but throughout the day.' But even 170 years ago, Thompson reported that, 'this bird suffers sadly during the mowing of our meadows. If the young have recently "come out" they are often either maimed or destroyed by the scythe.'[5] By 2022 there were less than 200 calling corncrakes recorded in the entire country with the majority found on offshore islands such as Tory in County Donegal where farming has all but been abandoned. Now the corncrake is on the brink of extinction in Ireland. We treat this as normal, forgetting that this bird was once a common feature of farmland in Ireland. Very few people alive today have ever seen or heard a corncrake. Each generation over the last two centuries thus has a different understanding of what is the 'normal' situation in the countryside. Our memories are limited to the human lifespan, and we find it hard to take into account what this same landscape would have looked like just 200 years ago. The baseline is constantly shifting, and we forget what has gone before.

The marine environment is no different when it comes to shifting baseline syndrome. Native flat oysters, which today are treated as a luxury food because of their rarity in the wild, were once so common that they were sold to poor people in the streets and exported in vast quantities to the markets of London. Overexploitation in the 17th and 18th centuries led to exhaustion of many natural oyster beds and these were later replaced by cultivation of imported Pacific oysters. If you visited any fishing harbour in Ireland in the early 19th century you would have seen hundreds of small fishing smacks landing huge quantities of herring, pilchards and other shoaling fish. An Irish fisheries scientist, John Molloy, wrote how many rural communities in Ireland at this time, particularly those along the more remote regions of the west and northwest coast, were greatly dependent on herring as food. Herring and potatoes were considered a staple diet and small farmers thought themselves inadequately prepared for winter unless they had a few barrels of salted herring stored away.[6] Once again, the temptation to exploit this 'endless' resource was irresistible and overfishing caused the collapse of many once-common stocks. We have forgotten what an important resource herring were for coastal communities.

This comes back to the question of what we consider to be the 'wild' state of nature in Ireland and what we are aiming to achieve through any restoration project. Perhaps there is a need to accept that nature can never return to a past condition that is long gone. If we can nudge it in the direction of greater biodiversity and stability, rather than trying to recreate the past, this will be achievement enough, and will be ensuring a better future for people and the landscape. In 2022, the European Commission published its proposed Nature Restoration Law and this was finally passed by the Council of Ministers in June 2024. This was a historic decision which sets in train a new initiative to restore damaged habitats and species. The

Irish government has already committed to preparing a Nature Restoration Plan by 2026. Every Member State of the European Union will have two years to publish a National Restoration Plan and set out how it aims to achieve the restoration targets that will be enshrined in law. EU Member States must then restore 30 per cent of the habitat types listed within the law to good condition by 2030, rising to 60 per cent by 2040 and 90 per cent by 2050. This is a positive move although it does not go much further than setting targets. It acknowledges that nature conservation on its own is no longer sufficient and that active restoration is required. While existing laws, such as the 1992 EU Habitats Directive, have been of value in preventing damage to certain internationally important areas, many of these places were already degraded and represent just a string of small fragments of nature that need active management to survive. In the marine environment, many of the Special Areas of Conservation (SACs) and Special Protection Areas (SPAs) are overfished or damaged by aquaculture. Drawing lines on a map and publishing 'conservation objectives' is not enough. Do we have to accept that a few 'semi-natural' but isolated sites are all that will remain in an impoverished general landscape or overexploited areas of the sea? At least the new restoration law will give some impetus to restoring these precious places.

Nature is not separate from human existence. It is not a luxury to be switched on like a television to entertain us. We are an integral part of nature, completely dependent on it for our present and future survival. We need fresh air, fertile soils, healthy vegetation and clean water. We need wild plants from which to breed our crops and we need insects to pollinate the plants that we eat. We need natural places such as the coast, mountains, lakes and woodlands for recreation and for general wellbeing. Without nature, we have no hope and no future. Despite decades of disappointing results from half a century of efforts in nature conservation, I remain optimistic. Nature-based solutions, such as rewetting peatlands, expanding forest cover or restoring seagrass meadows, offer some of the best hopes for averting climate catastrophe, the greatest challenge of our times. Here the objectives of climate action and nature restoration converge. In Ireland, there are already a number of nature-based solutions that are designed to restore habitats and species, but these are quite limited in time and space. This book investigates these innovative projects. Our current policies and practice in protecting nature are clearly inadequate to prevent the current loss and damage to biodiversity. Restoration is the urgent need now. The United Nations has declared the 2020s the Decade on Ecosystem Restoration which aims to prevent, halt and reverse the degradation of ecosystems on every continent and in every ocean. These high ambitions should spur more action to try to rescue nature across the world. But can the global restoration movement find support in this country?

This will not happen by itself. It will take vision, commitment and determination. It will require collaboration and buy-in by politicians, public bodies, commercial entities, voluntary organisations, private landowners and members of the public. But the prize is enormous. A healthy, self-sustaining environment, rich in wildlife, with real and permanent solutions to the biodiversity and climate crises. In this book I explore the ways that this can be done and talk to the experts, the people working to help nature recover. I pay tribute to those who dedicate their lives to this objective and the path that they have chosen. They demonstrate that wild places and species *can* be reintegrated into our landscape with benefits that all can enjoy. I don't want to minimise the problem of biodiversity loss or pretend that it will be solved by a few people doing their best. Successive Irish governments have abjectly failed to grasp the importance of the threats we face. In 2019, a climate and biodiversity emergency was declared in this country. But this has made little real difference to the sense of urgency. There have always been committed people swimming against the tide, working to protect and now to restore the natural environment, but they are far too few, finance is very limited and the scale of their projects is tiny.

Their work shows us innovative ways in which it can be done so this book is primarily about the *how* of restoration rather than the *why*. It does not shy away from the threats to nature but, primarily, it is about solutions. It is also about the people who are working to restore the country that they love. It is time to ramp up the action to restore nature to our landscapes.

In search of wilderness

When I was young, I read avidly about explorers in the wilder places on Earth – tropical Africa, the Amazonian jungles and the polar regions. I believed that there were still unexplored areas of the world that were so remote they were untouched by modern society, by technology or exploitation. I was wrong. Today, I know that human impacts have reached into every corner of the globe. Habitat destruction, species extinction, overexploitation of resources, air and water pollution and now climate change have all brought catastrophic losses of biodiversity and simplification of ecosystem processes. Is there anywhere that true wilderness still survives today? What does it look like? And can it provide us with the inspiration that we need to try to restore some of the richness in the natural world?

Throughout most of western history, wilderness was considered something negative, a dangerous place and a moral opposite to the realm of culture and godly life. In the colonial period, when Europe was thought of as the centre of civilisation, wilderness was viewed as being evil and resistant to being 'civilised'. The puritanical view of wilderness meant that, in order for colonists to be able to live in foreign lands, they had to destroy the wilderness to make way for their 'civilised' society. Today we have a more tolerant view, largely because there are virtually no unexplored parts of the planet left and the impacts of humanity are to be found almost everywhere. I have been fortunate during my life to experience nature in the wild, to draw inspiration and joy from simple natural phenomena as I return again and again to the open spaces that I love. Many of these places are highly modified versions of true wilderness. Tidal areas trapped within the walls of a port, woodlands that were planted to replace ancient forests or hillsides stripped of their natural vegetation by burning and overgrazing. I strive to understand the forces that power the natural world and the complexity which connects all living things to one another. The value of nature cannot be measured by any metric that we know.

In the early 1980s, I accepted an invitation to visit the unique Coto Doñana, in the south-west corner of Spain. I had read the classic book *Portrait of a Wilderness* by Guy Mountfort and was excited to experience at first hand a place that was apparently little affected by modern pressures. Mountfort wrote in the 1950s that there were still

a very few, small areas of wilderness remaining in Europe where nature reigns unchallenged in all her splendour. No roads lead to the Coto. To reach its fastness one must travel, as did the noble Dukes of Medina Sidonia on their hunting parties in the fifteenth century, by sailing up the broad Rio Guadalquiver and then riding on horseback for five or six hours, through the woods, across the desert and along the boggy edges of the great marismas.[7]

It sounded fascinating and my expectations were high for perhaps a last opportunity to see a great wilderness with such special animals as the lynx and the imperial eagle in the wild. For several hours, we travelled in a jeep with our guide through expansive dunes and pine woods, across the marshes and plains, past birds that appeared to show no fear due to the infrequency with which they saw people in this heavily restricted area. Once a small family of wild boar crossed our path, the tiny striped piglets apparently unafraid of the strange vehicle and its occupants. Discussing conservation issues with the biologists in the research station in the old *Palacio*, I learnt that the Coto is not immune from the pressures around it. Principal among these were the frequent shortages of water feeding the marshes as a result of increasing water abstraction for agriculture in the intensively farmed hinterland. Even more insidious was the effect of pesticides sprayed on rice crops upstream which found their way into the marshes via the river. Despite its apparent remoteness and rich biodiversity, Coto Doñana was a victim of increasingly intensive land use in the surrounding landscape.

The same process happened throughout Europe so that, by the 20th century, just a few small patches of original landscape remained. Białowieża Forest, on the border between Belarus and Poland, is one of the largest and last remaining fragments of an immense primaeval forest that once stretched across lowland Europe. It is probably best known as the habitat of the last surviving group of European bison, massive animals that still live wild in the forest. Covering an area of over 1,400 square kilometres (almost the size of County Leitrim), the Białowieża Forest World Heritage Site is seen by many as a remnant of what much of Europe once looked like. But it is far from untouched by humanity. It was a hunting reserve in the days of the Czars and, by the end of World War I, the original herd of bison had been wiped out, to be replaced later by the reintroduction of captive animals. The forest still retains some ancient oak trees, a few of which are estimated to be over 500 years old, and there is a high proportion of dead and decaying wood, a valuable habitat for forest organisms. However, despite numerous protective designations, illegal logging still threatens parts of the forest today. The Polish government has ignored pleas from UNESCO to stop logging in the old-growth parts of the forest, as well as an order from the European Court of Justice to halt the logging activities. Nowhere, it seems, is unique enough to be protected from human impacts.

At the northern end of the globe, another trip in the 1980s, this time to the wild landscape of East Greenland, helped me to understand what it felt like to explore a near wilderness. My first sight of it was from a big helicopter that skimmed across snow-capped mountains and valleys, its thudding engine the only human sound in a pristine landscape. As it descended to the floor of a wide valley I could see a vast river, divided into multiple channels. Finally, it settled on the ground and the four members of our team piled out onto the tundra, with tents, equipment and large boxes of provisions to last for a five-week period. After months of fundraising, planning and training, we had finally arrived at our destination in a remote valley in the arctic. Watching the chopper lift off and disappear over the horizon, I was filled with a mixture of excitement and loneliness. There was a stark realisation that I was now completely dependent on my three companions and our own resources for survival in a wild and deserted place, far from help in the event of an accident. Our team was there to study the breeding of barnacle geese which had just arrived from their wintering grounds in Ireland and Scotland. These are handsome birds with black and grey plumage, white faces and a haunting call that evokes the very wildness in which they raise their families in the short arctic summer. I had never before experienced the feeling of wilderness, a place so lacking in human impacts that it was as close to a natural ecosystem as any that existed on the planet. Occasionally, fur trappers passed through the valley, but there were no settlements, no farms, no trees, no roads and definitely no vehicles. Only the sounds of nature filled my senses. The incessant wind from the frozen sea close by, the sound of the river rushing past with its load of melting ice and the calls of birds echoed across the valley. But there was ominous news: prospecting for oil and minerals was already underway in the region and this could bring a whole new set of problems for this beautiful landscape.

Ten thousand years ago, as the last Ice Age glaciers retreated from Ireland, the landscape here probably resembled that valley in East Greenland, a true wilderness, still without a human presence. About a thousand years later the first settlers arrived to find an island covered with forests somewhat similar to the Białowieża Forest in eastern Europe today. These people spread throughout the island, pursuing a hunter-gatherer lifestyle. The advent of farming at the start of the Neolithic period 6,000 years ago brought with it a burst of forest clearance in Ireland. Thus began the conversion of wilderness into a human-dominated landscape. This story is eloquently told in Frank Mitchell's classic book *The Irish Landscape*.[8] Using the amazing science of pollen analysis, he charted the rise and fall of the vegetation and linked this with the various pulses of human activity across the land. Things changed slowly as the ancient woodlands were replaced by grassland and cultivated soil. The original wilderness, and all the species that it supported, were squeezed into smaller and smaller areas.

Does any wilderness still exist in Ireland today? Are there any sizeable areas that still retain their original complement of habitats and species? In the 1930s, the eminent naturalist Robert Lloyd Praeger wrote in his classic book *The Way That I Went*:

The Nephinbeg range of mountains is I think the very loneliest place in this country, for the hills themselves are encircled by the vast area of trackless bog. Where else even in Ireland will you find 200 square miles which is houseless and roadless – nothing but brown heather spreading as far as you can see, and rising along a kind of central backbone into the high bare hills breaking down here and there in rocky scarps, with the Atlantic winds singing along their slopes?[9]

Praeger qualified his comment by referring to the Nephin area not as depressing but inspiring. 'You are thrown at the same time back upon yourself and forward against the mystery and majesty of nature'. This idyllic picture may have changed somewhat today but this was the area selected by the state forestry company Coillte as the first 'wilderness area' in Ireland, later to be renamed as part of the Wild Nephin National Park. First unveiled at a conference in 2013, the project has undergone a number of changes but remains active a decade later. Coillte and the National Parks and Wildlife Service (NPWS) signed an agreement to set aside a total of approximately 15,600 hectares, as Ireland's first candidate wilderness area, to be known as Wild Nephin. This embraces two distinct areas – 11,000 hectares of undeveloped blanket bog and mountain to the west and 4,600 hectares of the Nephin Forest further east. The entire Nephin mountain range is included in the Park, stretching from Bellacorrick in the north to Mullaranny on the shores of Clew Bay. The forest part is the most significantly modified, by the establishment of conifer plantations on blanket bog and heath. Coillte's original plan was to take this area of non-native lodgepole pine and Sitka spruce out of commercial production, and to 're-wild' the plantation into a large-scale mosaic of mixed woods and blanket bog. Coillte said this would be achieved by thinning out the dense conifer stands, introducing native trees, and blocking forest drains to restore bogland. This would be a long-term project, addressed in stages.

Five years later a 'Conversion Plan' was commissioned to cover the period 2018 to 2033. By this time the whole area was under sole management by the NPWS, based on a long-term leasing arrangement. The declared objective was to 'facilitate the rewilding of landscapes and habitats in Wild Nephin to allow natural processes to become the dominant drivers in the landscape, where biodiversity is enhanced, ecosystems are restored and people can engage with nature through recreation for an authentic wilderness experience'. It confidently predicted that 'Wild Nephin will provide a unique and iconic wilderness experience of scale in Ireland, with opportunities for the enjoyment of nature, solitude and challenging recreation, without significant human presence or inappropriate activities, while supporting a wild and diverse ecosystem.'[10]

I decided to see for myself how this plan was being put into practice and so, on a wet autumn day, I joined Denis Strong and Sam Birch of the NPWS to tour the area. With thick mist hanging over the Nephin Mountains, I could see huge swathes of conifer plantation filling the lower ground. We began our walk at the start of the Letterkeen trailhead in a small stone bothy named in honour of the famous naturalist Praeger. All around this reconstructed building were thickets of the alien shrub rhododendron and occasional mature pine trees. Some small patches of rhododendron have been cleared by the NPWS, fenced to exclude sheep and deer and planted with native trees such as Scots pine, rowan and oak. These trial areas show what needs to be done on a much larger scale in the wider National Park. Sheep are still widespread on the commonage and, together with red deer that have escaped from captivity, they browse all the vegetation including tree saplings. Under the new ACRES agri-environment scheme, some farmers are being paid to fence areas of bog and reduce stocking rates with evident recovery of the vegetation.

In 2006, Coillte made a start on restoration of some of the wetter blanket bog to the north-east of the area by removing the lodgepole pine plantations in the townland of Derra. The wet conditions meant that it was impossible to fully remove the timber, so this was left to rot in piles known as windrows. Recent bird surveys have shown that red grouse are already returning to breed here. Coillte had planned to harvest large stands of conifers, but this operation was abandoned in 2021 and the whole area was handed over to the NPWS and rebranded as Wild Nephin National Park. Significantly, Coillte's Western Peatland Forests Project had already considered the redesign of forests along the western seaboard to develop a more sustainable land use policy for afforested peatlands, recognising that they were originally designed without regard for economic limitations or the ecological sensitivity of the receiving environment. The project concluded that the forests in question were largely uneconomic to manage due to their low productive capacity.[11] This implied that growing timber on exposed western blanket bog had been a mistake and should not have been undertaken in the first place.

Several important watercourses, such as the Owenduff and Bawnduff rivers, rise in the Nephin range and some of these hold valuable populations of salmon and sea trout which should benefit by removal of the conifer plantations. Freshwater pearl mussels have recently been found in the Deel River, making it vitally important to prevent pollution which could threaten this endangered species. Removal or thinning of the huge conifer plantations and control of spreading rhododendron alone present enormous challenges and will take many decades to achieve while ongoing maintenance is vital to ensure recovery of the blanket bogs. There is likely to be some establishment of small patches of native woodland within thinned areas of forestry wherever sheep and deer grazing can be excluded. The Conversion Plan estimates that this will take a budget of at least €30 million over the first 15 years and a large number of trained workers will be required in the difficult conditions. The original aspiration for a wilderness area seems as far away as ever but at least its National Park status may help this area to remain free of any new damaging developments such as roads or windfarms. Established land use practices such as forestry and sheep grazing, together with invasive species, remain the ongoing problems.

The European Commission has produced guidance on the management of wilderness areas in designated sites of European importance. Its definition states,

> A wilderness is an area governed by natural processes. It is composed of native habitats and species, and large enough for the effective ecological functioning of natural processes. It is unmodified or only slightly modified and without intrusive or extractive human activity, settlements, infrastructure or visual disturbance.[12]

It is clear that true wilderness no longer exists in Ireland. This country is one of the most heavily modified in Europe and there is very little left of the original habitats. Only a few areas can really claim to have suffered no serious impacts from human activity. The rocky shores that lie between high and low tide marks are relatively natural still but increasing sea temperatures and acidification due to climate change are likely to affect the plants and animals that live on them. High sea cliffs are generally inaccessible although the seabirds that nest on them are under significant pressure due to impacts on their food supply. Even the highest mountain summits are impacted by erosion due to overgrazing although fragments of natural habitat exist on cliffs and crags.

Even so, there are still places rarely used by people, far from traffic noise and polluted air, where a *feeling* of wilderness is possible. On the blanket bogs of the western counties, I can find a sanctuary far from roads, power lines and machines, with wonderful wild plants all around. At low tide, I can walk far out on sandflats in some of the larger estuaries, surrounded only by the sound of the waves and the calls of wild birds that spend part of their lives in the arctic. Sailing out on the Atlantic Ocean there are few other boats and I have spent many days with only seabirds and dolphins for company.

The American naturalist Aldo Leopold wrote, 'Wilderness is the raw material out of which man has hammered the artifact called civilization.'[13] In Ireland, wilderness is unlikely to ever return while our human population continues to grow and intensive land uses become even more widespread. Instead, we must look forward to a time when restoration work starts to regenerate some of our landscapes, leaving space for nature to live alongside people. Then, more and more people might be able to experience that raw *feeling* of wilderness.

Has nature conservation failed?

When I started my first salaried employment half a century ago, I was appointed as one of the wardens of a nature reserve in Northern Ireland. For a young naturalist, recently graduated with a degree in natural sciences, this seemed like a dream job. I could follow my passion for the natural world, spending most of my time outdoors studying plants and animals while at the same time working professionally to ensure the protection and enhancement of nature in this small patch of my native country. Life was idyllic. I lived in an old house on the end of a long line of sand dunes, just a stone's throw from a beautiful beach where I could wander at will, birdwatching, studying seals or exploring in rock pools. From colleagues, I quickly learnt about the business of nature conservation.

I grew up in Dublin, but it should be remembered that most Irish people in the mid-20th century were still country dwellers and, of those that lived in towns and cities, many were just a generation or two removed from their rural roots. Ireland remained largely dependent on agriculture in the early 20th century and farming in this country had not been modernised as in other Western European countries after World War II. Small, mixed holdings of 50 acres or less would have then been typical of much of the midlands, south and west of Ireland. The usual livestock on an Irish farm would have been a few cattle, some sheep on the hills and pigs in the backyard to fatten on scraps. Potatoes were almost universally grown to feed the large families and poultry would be kept for eggs and meat. Horses or donkeys were used to do the heavy work of ploughing, drawing turf (peat) from the bogs for winter fuel and bringing a little surplus milk and cheese to the dairy. Tractors and oil-powered machinery did not appear on Irish farms in any numbers until the 1960s. These relatively old-fashioned farming practices up to the 1950s also meant that there was little or no cash for the new imported fertilisers and farmyard manure remained the main source of nutrients that farmers used to maintain the fertility of the soil. So too, the chemical revolution of pesticides such as DDT, dieldrin and other organochlorines was not widely adopted in Ireland where farming was mainly grassland-based.

This delay in modernisation had several advantages for wildlife in Ireland. Firstly, the relative absence of persistent pesticides in the Irish environment allowed top predators such as otter and peregrine falcon to hang on here while they underwent catastrophic declines in other parts of Western Europe. Otters virtually disappeared in England during the 1960s and 1970s but were still common and widespread in Ireland as shown in a survey carried out in 1981.[14] While other Western European countries were undergoing massive losses of traditional farmland habitats such as hedgerows and species-rich meadows, Ireland managed to retain some of these habitats in its patchwork landscape of small fields. With these habitats, some of the more sensitive animals of farmland such as bats and corncrakes managed to survive in Ireland while declining elsewhere.

In Britain and Northern Ireland, although there had been voluntary bodies such as the National Trust and the Royal Society for the Protection of Birds since the late 19th century, nature conservation activity officially began shortly after World War II when attention turned to protecting the countryside that the British population had come to value so highly during the years of conflict. Thanks to a small corps of scientists and naturalists, the British government established the Nature Conservancy as a semi-autonomous body with its own research programme, land ownership and advisory roles.[15] While this meant that the protection of important sites got off to an early start in the UK, it is debatable whether this approach has been in any more successful there than in any other country in Europe.

The much later state conservation model in Ireland was quite different. The Forest & Wildlife Service was set up as part of the larger Department of Lands, at the bottom of the civil service pecking order. In the 1960s all the attention of government was on growth in an economy that was, until then, heavily dependent on agriculture. At that time, anyone calling for the protection of threatened habitats or species was generally associated with a privileged elite and received minimal support from government. In a debate on the Wild Birds Protection Act in 1930, the politician and later Taoiseach Seán Lemass TD said in the Dáil, 'we must put the necessities of human beings before those of wild birds'.[16] Attitudes to nature conservation in Ireland are still much influenced by the recent history of an island where land ownership was only attained by many Irish people at the start of the 20th century and the rights of landowners are very strongly protected in the constitution. This has proved to be a difficult issue for nature conservation especially in persuading private landowners that nature should be protected on their lands.

In the 1940s the Arterial Drainage Act was passed into law allowing the Office of Public Works to carry out untold damage to rivers and wetlands throughout the country. Between 1948 and 1995 the Office of Public Works (OPW) completed 34 arterial drainage schemes and five estuarine embankment schemes amounting to over 11,000 kilometres of river channel. This dried out a total area of land equivalent to the size of County Roscommon. Most of the natural features of rivers, such as meanders, riffles, pools and gravel substrates, were removed leaving them more like straight, steep-sided canals. Major casualties were some of the best salmon spawning rivers in the country with unknown numbers of other freshwater species as collateral damage. One example was the freshwater pearl mussel, which was already suffering from loss of water quality in Irish rivers, and which depended on salmon and trout for survival during its life cycle. The loss of the Irish raised bogs had continued apace from the formation of Bord Na Móna (the Irish Turf Board) in 1945. This virtual obliteration of the peatlands for fuel and garden compost, again by a state-sponsored body, later became the focus of an international campaign to save these unique habitats.

With modernisation of the country well underway in the early 1970s, Ireland was negotiating for entry to the European Economic Community (EEC) – an appropriate time to demonstrate that we too were good Europeans. The European Conservation Year 1970 campaign brought many public events and new publications, while media coverage of conservation issues began to appear for the first time. Some Irish people, at least, began to take note that wildlife was more than simply game to be hunted or vermin to be controlled. In February 1970, a remarkable meeting took place in Killarney, County Kerry. It was a conference entitled 'The Future of Irish Wildlife' organised by the Agricultural Institute (An Foras Talúntais). A glance at the list of participants given in the published proceedings shows that this meeting included a glittering cast of the leading international conservationists of the time. Most of the Irish participants were from government departments, state research institutes and universities.

Notably, Irish non-government organisations were not formally represented.[17] The conference made a number of recommendations, but these mainly concerned the need for more research and did not seriously address the new pressures on the natural environment. Voluntary nature organisations were just getting going at this stage with the Irish Wildbird Conservancy (now Birdwatch Ireland) formed in 1969. Accession to the EEC (now European Union) in 1973 brought Ireland into alignment with other European nations and started the process of implementing nature conservation at an official level.[18] However, a real nature conservation agenda only took legal effect from 1979 with the adoption of the EU Birds Directive and the later Habitats Directive. Despite signing up to these laws, Ireland's extremely poor record on implementing them showed that they were not taken too seriously by the authorities here.

In the 1970s, most of the effort was devoted to surveys to try to document what remained of nature and the listing of 'areas of scientific interest'. Legally protected areas were limited to nature reserves on state-owned land, the early sites having been originally purchased for planting with fast-growing coniferous forests. The Wildlife Act of 1976 gave legal protection to a short list of birds and mammals as well as limiting the hunting of certain quarry species to open seasons. However, the wider countryside, outside of nature reserves and national parks, and most other wild species, received no legal protection. I started working for the National Trust in Northern Ireland in

1974 and my own focus at that time was the management
of a single nature reserve on the coast. The approach
of conservationists then was to put a fence around any
protected habitat and keep people out as they were seen
as causing damage through disturbance, erosion and risk
of fires. The need for habitat management to try to restore
degraded habitats was in its infancy.

By the 1990s, things began to change with the passing
of the EU Habitats Directive which required not only
legal designation of internationally important areas but
conservation objectives to be established and enforced.
Even then, the emphasis in Ireland was on preventing
further loss of these areas with little attempt to restore any
damaged habitats or reintroduce lost species. It should
be noted that this is not the only objective of either the
EU Birds or Habitats Directives, both of which require
restoration where species and habitats are not in good
conservation status. However, the majority of designated
sites are privately owned in Ireland in comparison to other
EU countries, which have a much greater proportion
of state-owned land in protected areas. This presented
significant challenges to working in partnership with Irish
landowners which has still had limited success to date.
The rest of the countryside outside of EU-designated areas
was largely left for 'business as usual' with few controls
on habitat loss or damage in farmland, mountains, bogs,
coastline, wetlands, rivers or other semi-wild areas.

The early 2000s were a depressing time for nature
conservationists as development pressures accelerated

during the 'Celtic Tiger' period and nature took a poor second place. Motorways and housing developments mushroomed across the country, some threatening important natural habitats and wildlife. Agriculture received substantial supports from the EU to modernise resulting in widespread intensification, greater chemical use, removal of hedgerows and increasing water pollution. Nature was considered an obstacle to development. The Taoiseach of the time, Bertie Ahern, told the Dáil that opposition to motorways was about 'swans, snails and people hanging out of trees'. Ireland was charged in the European Court of Justice for its failure to protect important habitats such as peatlands from exploitation. Following the recession of the 2010s development pressures stalled, but this also brought cuts in public funding for nature conservation which had major implications for the capacity of the official conservation body, the National Parks and Wildlife Service, to do its job properly. The NPWS continued to be moved around between government departments as if no politician wanted to take responsibility for this difficult issue. So, the short history of nature conservation in Ireland is one of apathy, inconsistency and lack of political commitment.

What was happening to nature in the face of this apathy and pressure over the last half-century? Headlines include significant losses of habitats and species. Raised bogs were virtually erased, non-native forestry planted over peatlands and high nature value farmland, uplands burned and overgrazed, ground-nesting birds seriously impacted, insect populations decimated by habitat loss, river quality damaged with migratory fish populations suffering and marine resources increasingly overfished. Threatened species such as the corncrake, red squirrel and Atlantic salmon received much attention but many other less charismatic species such as insects, fungi or lichens were largely ignored in official circles. The journalist George Monbiot wrote that, 'Conservationists sometimes resemble gamekeepers: they regard some of our native species as good and worthy of preservation, others as bad and in need of control.'[19]

This catalogue of destruction and loss of biodiversity suggests that nature conservation has been failing for a long time or, at least, not achieving its targets. So, is there a future for conventional conservation methods in Ireland? Simply designating 'protected areas' and 'protected species' is not enough without active management work. The focus of conservation has largely been on scattered 'protected areas', while other habitats and species disappear, and the general environment continues to degrade. The most recent national assessment shows that the majority (85 per cent) of the internationally important habitats that are theoretically protected within SACs are in unfavourable (that is, inadequate or bad) status, with 46 per cent of habitats demonstrating ongoing downward trends. The NPWS clearly states that there are 'inadequate conservation measures in place to improve the future prospects of these habitats' and that 'declining trends are particularly notable in marine, peatland, grassland and woodland habitats'. The most frequent pressures recorded in these irreplaceable sites relate to agricultural practices, and the pressure is ranked as high in more than 50 per cent of the habitats. The commonest cause of damage is intensive grazing or overgrazing by livestock causing serious impacts on the value of these places for nature.[20] These assessments, important though they are to keep track of trends, seem to be simply charting the declines without doing anything to reverse them. While specific 'conservation objectives' have been published for many of the protected sites, there appears to be little effort to implement the necessary restoration or persuade private landowners to do so. Is this fiddling while Rome burns?

It must be said that there are some benefits to site designation which are virtually invisible to most people. Legal designation as SAC or SPA means that these sites must be taken into account by planning authorities when assessing proposed construction projects such as roads and housing. The European directives require that very detailed ecological studies (known as appropriate assessments in the business) must be carried out by the developer before permission can be granted. This has modified, held up and even caused abandonment of some large state projects such as motorways or the infilling of parts of estuaries for port development. But the best that can be hoped for is that they are constructed elsewhere in a less sensitive location. It could be said that the EU directives and the European designations can prevent some bad things occurring in the protected areas but they have not arrested the gradual decline in biodiversity in such areas, and have rarely caused any good things to happen to them. Correctly implemented, the Directives can certainly slow and reverse biodiversity decline in designated sites such as the Burren. The fact that there has been very limited success so far in Ireland, mostly linked with EU-funded LIFE projects, does not mean that it is not possible or feasible.

For most 'protected species' the trends are a little more encouraging. The most recent government assessment of those protected under EU legislation shows that the overall status of the some 60 regularly occurring protected species (including some species groups such as bats and whales) is that 57 per cent of species have a favourable status and 30 per cent are considered to be in unfavourable (that is inadequate or bad) condition, with 72 per cent demonstrating stable or improving trends and 15 per cent having ongoing declining trends. The species on the downward slide include invertebrates such as the freshwater pearl mussel, the freshwater crayfish and the lesser horseshoe bat. Birds are covered by separate European legislation and, for many of them, the Irish picture is not good. BirdWatch Ireland and the RSPB jointly prepare regular assessments for the whole island called the Birds of Conservation Concern in Ireland. The criteria on which these assessments are based include conservation status at global and European levels and within Ireland, historical decline, trends in population and range, rarity, localised distribution and international importance. The conservation status of species is signalled using a traffic light system. Of the 211 bird species covered in the most recent assessment of 2021, 26 per cent were placed on the Red list, 37 per cent on the Amber list and 37 per cent on the Green list. This means that over a quarter of Ireland's birds are declining with a number of these in danger of complete extinction on this island.[21]

Of course, these evaluations relate only to the rarest habitats and species on a European scale – places like the Burren or Bull Island in Dublin Bay and species like rare birds or bats. Only one insect species, the marsh fritillary butterfly, out of the tens of thousands known in Ireland, has legal protection under the EU Habitats Directive. Nor do the designated sites cover the whole range of most species. For example, protecting seabird breeding colonies while leaving their foraging areas subject to overfishing can undermine the whole marine ecosystem. There are, of course, many species and habitats that are not protected under EU legislation and most are not even protected under national legislation here. There are myriads of 'ordinary' species, far too numerous to count, that go unprotected in the Irish landscape and surrounding seas. What about the thousands of insect species, many of which are unstudied, common bird communities that are quietly losing habitats or some of the sensitive plant species whose remaining places are disappearing at an alarming rate? What about the dozens of marine species that are being overexploited by fishing fleets while the scientific limits on catches are flagrantly ignored? These mostly remain unprotected and continue to degrade or disappear at a rate almost unrecorded. Local communities and concerned conservationists often fight lonely battles to protect habitats and species against huge odds.

Species-rich grasslands are a good case in point. If a beach or a lake was destroyed there would be a public outcry. But if a flower-rich grassland was ploughed up would anyone even notice? This is the question posed by Dr Maria Long of the NPWS. 'Semi-natural grasslands', she says, 'are invisible.'[22] About 28 per cent of hay meadows and 31 per cent of calcareous grasslands, such as those in the Burren, have been lost in the last six years, according to a new survey by the NPWS. Corncrakes and breeding waders of grasslands, such as the curlew and lapwing, have also suffered catastrophic declines according to Birdwatch Ireland. So too have many species of grassland insects – butterflies, moths, bees and hoverflies – as reported by the National Biodiversity Data Centre.

National biodiversity action plans have been prepared by successive governments (the fourth one was published in early 2024) but these simply restate the problems and set targets for recovery without the necessary governance systems and financial/human resources being put in place at the same time to deliver these objectives. Government conservation agencies have been seriously under-resourced which has largely limited them to surveying the declines in Irish wildlife and designating 'protected sites', many of which were already seriously degraded. An independent review of the National Parks and Wildlife Service in 2021 concluded that, 'The NPWS cannot meet current obligations, let alone plan for and respond to future challenges and legislation, including the Climate Action Bill and EU Biodiversity Strategy to 2030.' It also noted that a number of EU environmental directives had not been fully implemented and multiple infringement cases were outstanding against Ireland. The review, prepared by Professor Jane Stout and Micheál Ó Cinnéide, found that,

> While NPWS staff are dedicated, passionate and knowledgeable, there needs to be a fundamental overhaul of structures and governance, a clear strategic plan and leadership to implement it, better internal and external communications, and re-energised teams, working together effectively inside and beyond the organisation.

This catalogue of destruction and loss of biodiversity in Ireland suggests that conventional nature conservation has been failing for a long time or, at least, not achieving its targets. So, is there a future for the old methods of nature conservation in Ireland? The efforts to defend 'protected areas' and 'protected species' must continue as their legal protection is an important bulwark against complete destruction, but these 'special' areas are largely isolated islands of nature in a sea of intensive land uses. Protected species are frequently struggling with seriously degraded or damaged habitats, badly polluted waters or competition with invasive species introduced from abroad. There is obviously an urgent need to alter current land use practices on a broad landscape scale. For example, intensive beef and dairy farming should be replaced with more extensive livestock rearing and organic crop production, drastically reducing the use of artificial fertilisers and chemical pesticides. The conventional practice of forestry in this country is damaging nature and is more like intensive crop production. There should be a widespread conversion to continuous cover methods, also known as close-to-nature forestry. This should have a much greater focus on the use of a diversity of native tree species instead of monocultures of non-native conifers. Overgrazing and burning of vegetation in the uplands, which prevent natural regeneration of trees, should be halted and the areas of damaged peatland restored. In the marine environment, the predominant threats are overfishing and damaging practices such as bottom trawling and dredging. These need to be urgently curtailed and the people who make a living from these practices given alternative employment or compensated to prevent the damage.

Even if the obvious damage could be limited or slowed down in a short timeframe, there would still be a need for active measures to restore the habitats and species that have been declining so rapidly. Such action has been tried in some areas already but many of these projects are little known. For example, small areas of native woodland are being planted on private land under government grant schemes. Some rewetting of bogs has been done in National Parks and Bord na Móna has started one of the largest peatland restoration projects in Europe on the midland raised bogs. A few charismatic birds of prey, such as the red kite, have been reintroduced to Ireland after absences of hundreds of years. But the success stories are few and far between. There are plenty of examples of large-scale restoration projects in other countries and other continents.[23] In Scotland, for example, which has similar habitats and pressures on nature, there are active projects to restore the native Caledonian pine forests, right across the country from east to west. We know what works. We just need to get on and do it.

Restoring the woods

As I wandered through the old oakwoods around the Lakes of Killarney, it was easy to imagine a prehistoric time when Ireland was covered with primaeval forests like this. A mist hung over the lakes and clung to the old trees. Dripping from the leaves, soft rain soaked the ground, a thick blanket of mosses covering the rocks. From somewhere high up the hill, a red deer stag roared to declare his territory.

There is a widely held perception that Ireland was originally covered from coast to coast with native forest. I have heard it said that, at one time, a squirrel could cross the country from one side to the other without ever touching the ground. Certainly, the fossil pollen record shows that a succession of tree species colonised Ireland after the last ice age as the climate warmed. First pine, yew and birch then oak, elm, hazel and ash followed by a diverse range of native species migrating north from the European continent. But there were also large areas covered by lakes and later peatlands that would have been largely free of trees while herbivores would have grazed out clearings and created grassland through repeated browsing of the undergrowth. So, we might envisage more of a patchwork of forest, grassland and water, intimately mixed in a varied natural landscape.

By the beginning of the Neolithic (Late Stone Age) period, the first farmers were already tilling the rich soils with crude stone implements and wooden ploughs pulled by oxen. The archaeological record is complemented by pollen records from nearby lake muds and peats which allow a reconstruction of the type of farming carried out and how this affected the woodlands. The oak woodlands would have provided plenty of foraging for domestic animals, particularly the early breeds of pig that foraged for acorns in the autumn and the underground storage organs of plants like bluebell and pignut during the rest of the year. As the undergrowth was grazed back, fewer young tree saplings survived and gaps opened in the canopy as older trees fell.

It is well known that 'the history of Irish woods is replete with uncertainties, misapprehensions and elements of paradox'.[24] The impact of Neolithic peoples on the primaeval woodlands from about 5,000 years ago is certain. If the forests had been dense then they would have had to create clearings for their domestic animals and crops. If it was a patchwork already, due to grazing wild animals, then the early farmers probably concentrated on the grasslands first and gradually expanded these. Dr Oliver Rackham of Cambridge University believed that in the uplands and the West of Ireland much of the wildwood was converted directly to moorland and that natural processes of climate change played a much greater role in the loss of the ancient forests here than in England. In the western areas of highest rainfall, woodland was replaced by the progressive accumulation of blanket peat. Upland areas may have been used for rough grazing and large burial cairns were built on the summits.[25]

Guesstimates have been offered that no more than 20 per cent of the island was still covered in forest or shrubland by 1000.[26] We had to wait until the medieval period for the first real written natural history of Ireland. Giraldus Cambrensis came to Ireland for the first time in 1183. He kept detailed diaries of his tours and when he returned to Wales he wrote, in Latin, his great book *Topographia Hiberniae (The History and Topography of Ireland)*. He described Ireland as 'a country of uneven surface and rather mountainous. Still there are, here and there, some fine plains, but in comparison with the woods they are indeed small.' Cambrensis reported the presence in Ireland of some typical woodland mammal species such as 'boar, marten and badger'. He reported that 'there are only two kinds of harmful beasts in Ireland, namely, wolves and foxes'.[27]

Although this is often disputed, one estimate has suggested that by 1600 about one-eighth (12.5 per cent) of Ireland was still covered with natural forest.[28] Ancient woodlands in Ireland today are defined as areas of forest believed to have remained continuously wooded since

1660. This date is used as the first accurate mapping took place in the Down Survey and the Civil Survey which were undertaken in the 1650s to facilitate the confiscation and reallocation of lands following the Cromwellian conquest. Shortly after this time, planting of new woodland by English landowners began to enhance their estates.[29] Small fragments of the original tree cover continued to decline or were replaced with mixed forest grown for its timber value or for coppicing to produce fuel for smelting iron. By 1800 only 2 per cent of the original native forest remained.

Today, just 11 per cent of Ireland is covered with all forests but the vast majority of this is made up of ranks of exotic conifers such as Sitka spruce, lodgepole pine and Douglas fir. Less than 2 per cent is comprised of native trees and the majority of this is in tiny patches of less than ten hectares, just fragments of the wide blanket of native forest that once covered the country.[30] Efforts to expand the area of native woodland in Ireland have been frustrated by a number of factors, some ecological, some administrative, others due to inertia in the forestry industry. Nevertheless, there is an increasing interest in planting trees and a whole range of voluntary organisations has been formed with this single objective. Grants are available from the government to support the initial costs of fencing, ground preparation, nursery plants and trained teams to undertake the planting and subsequent management.

One of the first things we considered after acquiring our own small farm in County Wicklow was to plant some of the pasture with native trees. After fencing off some fields to exclude livestock and deer, we got to work one winter with the help of teams of volunteers. Planting large numbers of oak, birch, Scots pine, hazel, wild cherry, crab apple and rowan saplings was a steep learning experience. Some trees suffered from summer drought, competition from rank grass and damage by rabbits and hares. I quickly learnt that tree-planting is just the first stage and that new woods require continual management to ensure success. Under the grant scheme, a landowner may receive an annual premium for up to 20 years to cover the time and cost of maintaining the plantation, to compensate for taking the land out of agriculture and to bridge the gap until there is some income from the sale of timber or firewood. However, there is still a reluctance among landowners to switch their land from farming to forestry or to change from the practice of planting fast-growing conifers such as Sitka spruce that can be harvested within a few decades by clear-felling, thus giving a quicker return on the investment. This has been the common approach to afforestation since the foundation of the state a century ago.

Nevertheless, since 2002 there have been many private landowners grant-aided under the Native Woodland Scheme to create new areas of native woodland or to restore older woodlands that are of high nature value. The grant-aid covers such actions as deer fencing, ground preparation, tree-planting and removal of invasive species. To date over 4,300 hectares of new woodland and 3,100 hectares of conservation in existing native woodland have been grant-aided in this way. To put this in context, even combined, this is only about one-third of the area covered by golf courses in Ireland. In 2023 a new programme of grants for landowners was launched by the Forest Service with enhanced grants supporting a wider range of options including agroforestry and close-to-nature forestry.

Zef Klinkenberg and Mandy Hollwey did not wait for a grant to start planting trees. When they bought a derelict cottage and a small, abandoned farm in Glendasan, County Wicklow, in 1989, they immediately began to clear bracken and gorse from the steep rocky land. The farm, which is adjacent to the Wicklow Mountains National Park, was overgrazed by herds of invasive sika deer browsing the vegetation so that no tree seedlings could survive. All newly planted trees had to be protected in plastic tree guards until they grew up beyond the browsing level. Across the picturesque Glendasan River is a stand of mature Scots pine trees that was planted by miners in the 1880s and here Zef began to add a variety of broadleaf species along the riverbank. Once the land was fenced, planting began in earnest. The steeper parts included a mature Sitka spruce plantation which was clear-felled and removed, leaving an area of bare, disturbed ground. Here, the couple planted several thousand young trees including oak, hazel, birch and rowan. Today this has developed into a young mixed woodland with occasional gnarled old birch trees that survived from an earlier forest. Zef says,

What surprised us was the huge number of native trees that naturally regenerated particularly in the disturbed ground on the clear fell site once the area was protected from the browsing of deer. This included more oak and birch but also lots of holly, Scots pine and willow.

Heather and a variety of wildflowers were able to flower in abundance in the wet mountainy land and birdlife returned. This clearly demonstrates the positive outcomes that come from the removal of conifers and protection from grazing. Thirty years later, Zef and Mandy can relax in their beautifully restored cottage while looking out on a recovering hillside that is rich and diverse.

Gilly Taylor and Brian O'Toole also had a vision for the future when they bought seven hectares of sheep-grazed pasture on the bank of the Redcross River, not far from the coast of County Wicklow. The land is very wet and, when they dug some large ponds, these filled immediately with water. With grant aid from the Native Woodland Scheme, they decided in 2020 to plant up almost 2.5 hectares of the wetter fields with 6,700 native trees, mainly alder but also a scattering of hazel, Scots pine, hawthorn, oak and birch. Despite some initial difficulties with a summer drought, damage to the young trees by hares and a dieback disease on some of the alder trees, the plantation is now successfully growing and merging with the mature trees in the hedgerows. Gilly says, 'As we did not want to use herbicides, vegetation control has been all done manually, initially by the forester, and in subsequent years by ourselves and a wonderful army of volunteers.'

Adding a whole suite of additional ponds and a sizeable wildflower meadow, the place is now called Wildacres Nature Reserve and has already attracted a wealth of wildlife including pine marten, hare, fox, badger, otter, kingfisher, sparrowhawk and woodpecker. Here the couple host a range of engaging and entertaining nature-based tours and interactive workshops for individuals, community groups, clubs and corporates.

One of the species planted at Wildacres was Scots pine, the only native conifer in Ireland. A mature Scots pine is very distinctive, tall and straight with a reddish-orange tinge to the bark. Due to the dense flat canopy, it also provides excellent nesting sites for some birds of prey such as sparrowhawk, merlin and red kite. For a long time, it was thought that Scots pine had died out completely here around 400 due to a combination of climate change and human activity. It is still a common tree in Ireland but most of the Scots pines we see today are derived from introductions from Scotland. However, a study of the pollen record from a unique stand of pines at Rockforest in the Burren showed clearly that, at least in that location, there was a continuous record of the species from the end of the last ice age to the present.[31] Now, a recent genetic study has shown that all the Irish Scots pines are likely part of a common Irish-Scottish gene pool.[32] This means that they form a single population and that even the introduced stock here is considered native. One of the authors, Dr Colin Kelleher of the National Botanic Gardens, says this is 'like welcoming back a long-lost relative when we are reintroducing them'. This should spur more planting of this attractive and valuable tree as we continue to expand the native woodlands of Ireland.

By contrast with tree-planting, natural succession is nature's way of returning an area of open ground to woodland over a period of decades. If intensive grazing can be excluded and if there are mature trees in the area that produce a seed source, then scrub and trees will establish by themselves. This process is described in *An Irish Atlantic Rainforest,* a popular book about one landowner's experience in the remote Beara Peninsula of West Cork. Eoghan Daltun acquired a semi-derelict farm and 30 hectares of land where livestock grazing had ceased several decades earlier. As a result, much of the land was covered with native forest, although this was in poor condition due to overgrazing by deer and wild goats as well as invasion by alien plants, all of which prevented natural regeneration of trees. By fencing off the woodland and tackling the invasive rhododendron he has managed to return this to a patch of Atlantic rainforest, rich in wild plants and animals. He believes that nature knows best which tree species are suited to the soils and drainage on a particular piece of land.[33] This special type of native woodland in the West of Ireland, now called Atlantic rainforest, is also the focus of a charitable organisation called Hometree. It has launched a new project which seeks to restore temperate rainforest on 2,000 acres of its own land and a similar area in collaboration with adjoining landowners. It will do this by fencing off remnant pockets of surviving forest, allowing natural regeneration to take place, with additional planting of native trees 'where there is a strong ecological rationale to do so'.

The same principles are being employed in some large areas in other countries in order to expand their native woodland. In the Southern Uplands of Scotland an entire valley has been restored to what is now called the Carrifran Wildwood. The 650-hectare property was acquired by the Borders Forest Trust in 2000. Most of the plant life at that time consisted of short-cropped, tree-less vegetation typical of long-term grazing and *muirburn*, the Scottish word for systematic burning to make the habitat more attractive to grouse. The ancient forests had long since gone and had been replaced by swards of mat-grass, bent and fescue grasses with some areas of heather. A few tiny areas of the old forest hung on in inaccessible places such as cliffs, ledges, steep ravines and wherever domestic animals and goats could not reach. The first stage was the removal of sheep, cattle and feral goats that had denuded the landscape in the previous millennium. The fragments of woodland were very small but there were about 19 native species of trees and shrubs which did provide a seed source to recolonise the areas released from grazing.[34]

In addition to natural regeneration of woodland, some 650,000 trees and shrubs have been planted at Carrifran since 2000. Local geology, soils, climate and existing vegetation were used as indicators to decide the most appropriate type of woodland that would be successful in each area. After two decades there was already a closed canopy, vertical and horizontal layering, leaf litter and decaying wood building up on the woodland floor. In short, the ecological functions of a forest were beginning to operate again. Where there had been a bare landscape stretching away to the summits there was now a developing forest filling the valley with a mix of tree species and many other native plants colonising by themselves. The change in a short period of time is dramatic and illustrates what is possible given the availability of land and a long-term commitment to nature restoration.

Glenveagh National Park is located in the centre of the Donegal uplands, in a landscape that resembles parts of the Scottish Highlands. Even the baronial-style castle by the picturesque lake is reminiscent of Victorian images from Scotland. Centuries ago, the slopes and valleys here were filled with ancient woodland with an area of 227 hectares estimated from old Ordnance Survey maps, although this area was not all wooded at the same time. Many of the townland names include the word *doire* which means oakwood in Irish. Unfortunately, these woods have been whittled away over the years due to lack of management, extensive deer grazing and shading from invasive species such as rhododendron and a range of other troublesome plants. Today, just 130 hectares of the original woodland remain, of which the majority (82%) is broadleaved woodland, mainly dominated by oak. In 2022 the government agency NPWS commissioned a woodland management strategy which aims to restore, expand and reconnect the woodlands of Glenveagh. It reviewed all the historical information and mapped the remaining woodland before proposing a series of actions to take forward.

The new draft strategy proposes a comprehensive management programme for controlling the invasive species to allow more light to penetrate the woodland floor. Some 45 kilometres of deer fencing were erected around the estate in the 19th century but this has been breached over the years and is no longer effective. The absence today of any natural apex predators of deer, such as wolves, lynx and bears, has led to a significant increase in the number of deer, which seriously damage the regeneration potential of the woodland through browsing young tree seedlings and stripping bark. Instead of repairing an ineffective fence, the new strategy recommends localised culling to reduce the deer densities and hence damage to the woodlands. Once this is achieved, deer densities then need to be maintained at the lower level through continued culling and monitoring. It is recommended that carcasses be left on the hill to provide carrion as a food source for invertebrates and fungi, as well as larger scavengers such as golden eagles, ravens, foxes and badgers. To enhance the woodland regeneration a tree nursery has been established at Glenveagh, using locally sourced seed to produce young trees which will subsequently be planted out with appropriate protection from deer. This will be complemented by selective removal of non-native trees and thinning of the canopy to allow more sunlight to reach the woodland floor. Together with comprehensive monitoring and a public information programme, these measures should ensure a better future for the valuable native woodland at Glenveagh. Dr Jenni Roche, Restoration Planning Ecologist with the NPWS, says, 'Full implementation of this ambitious plan will ensure that the internationally important woodlands of Glenveagh can recover and flourish in future. We hope it will provide a model for other landscape-scale restoration projects in Ireland.'

In the east of Ireland, in particular, a number of broadleaf woodlands were managed historically by the ancient practice of coppicing. This is the act of cutting down trees to induce regrowth while the regrowth itself is termed a coppice, or *coupe*. All deciduous trees can be coppiced and will regrow from the base while most native species respond in this way, with the exception of Scots pine. There is an old saying in north-west England, 'If you want wood, you have to cut wood.' This is a form of nature restoration as it ensures that the woodland is a permanent feature and the management helps to restore typical woodland flora with the return of sunlight to the woodland floor. The largest area of coppice was once in County Wicklow where up to one-fifth of the county was owned by the Fitzwilliam family on the Coolatin Estate and managed in this way. Large numbers of woodworkers were employed using only hand tools and horses to cut, manage and extract the timber which was divided into mature 'standards' and short rotation 'underwood'. Typical species that were managed in this way included oak standards and hazel underwood, each managed on a rotational basis with different periods of regrowth.[35]

Mike Carswell is an experienced coppice worker and joiner from County Down. He trained as an apprentice in England and subsequently moved his coppicing business to Ireland where he now manages a large plantation at Carrigmore near Wicklow Town. I walked around the woodland with Mike on a fine spring day and I was struck by the systematic way in which he is restoring this site on a gradual basis, at the same time producing a wide range of useful products such as building spars, hedge-laying stakes, bean poles and charcoal. Birdsong filled the wood and the ground was bright with wildflowers such as wood anemones and primroses. It is hard physical work, but rewarding to see the rich biodiversity develop and the woodland restored. Recently, Mike has begun to pass on his extensive knowledge through a series of training courses sponsored by the government Forest Service.

In Ireland, the state-owned forestry company Coillte is one of the largest landowners in Ireland with an estate that covers approximately seven per cent of the country. Less well known is the fact that about one-fifth of this is managed with biodiversity as the main objective. There is quite a bit of unplanted or 'unplantable' land, but some is mixed old woodland that was bought from large estates around the country. Recently, the company has set up a not-for-profit entity called Coillte Nature that is dedicated to the restoration, regeneration and rehabilitation of nature across Ireland. Its stated mission is to deliver real impact on the climate and biodiversity crises through large-scale projects based on the best ecological evidence. I met with the Director of Coillte Nature, Dr Ciaran Fallon, to find out exactly how his team is setting about this challenging task. Over the short few years since its establishment, Coillte Nature has taken on a range of projects, some on existing parts of the company estate and others in collaboration with public bodies like Bord na Móna. Ciaran believes that, in the past, there have been too many projects which have just been test studies or proof of concept. He says, 'We are beyond that now. We have to step into it in a purposeful and deliberate way with a sense of urgency.'

He also knows that the projects need to be scaled up to make a significant impact on the loss of biodiversity. He says, 'We have one project that is funding the restoration of 2,100 hectares in the western peatlands. This is currently coniferous forestry planted in the 1950s and 1960s, which was, at the time, a social project to try to keep people in rural areas.' One problem now is the absence of people to do the necessary restoration work.

> For example, we have a very large project in Derryclare, County Galway. All the practical work is done by either contractors or by colleagues in the commercial forestry operation who have the resources. I believe we can get to a stage where we can be rehabilitating around 1,000 hectares of

degraded peatlands every year, but to do this we will need to develop the skills and capacity of local contractors to work at scale on these sensitive landscapes. We need to create a small army to do the actual restoration work.

A lot of the work involves removing forestry that was originally planted in the wrong places, on peatlands where they did not grow properly because of waterlogging and lack of nutrients. Ciaran says,

It tends to be North American species such as Sitka spruce and lodgepole pine. These peatlands were planted with commercial conifers to create rural employment and it involved massive amounts of labour, people digging drains, planting trees and applying large amounts of fertiliser largely by hand. At the time carbon emissions and biodiversity were simply not a consideration. In hindsight we now know that this was not a good idea for climate, nature or forestry. These peatlands are not suited to growing commercial trees. In areas where it is impossible to restore the growing peatland, we will work with native Irish trees such as birch and alder to develop a bog forest.

In some areas, invasive species are a key issue. At Hazelwood in County Sligo, rhododendron is a big problem. Ciaran says, 'the forest here is very difficult to work in due to flooding but we are gradually removing the rhododendron and restoring the site.' On the midland cutaway bogs, Coillte Nature is working with Bord na Móna, on their lands, to find out what techniques will work to establish native woodland where there is no potential for recreating a functioning peatland. 'The trials currently involve spreading tree seed on about 200 hectares and then expanding that out to 1,500 hectares.'

Coillte Nature has also set up The Nature Trust which has separate finances and is a company limited by guarantee. The Trust is committed to planting new, native woodlands and revitalising natural wild habitats on a large scale across Ireland. It manages these lands on a non-commercial basis for the benefit of ecosystems, the climate, local communities and wider society, now and for generations to come. Its finance comes largely from the business community. For example, an insurance company is providing five million euro to support the planting of over 400 hectares of new native woodland across Ireland. This is often referred to as Corporate Social Responsibility or 'offsetting', if it provides a benefit to the company. A typical site in this programme is at Aughrim, County Roscommon, which is a scenic location. There is a stream running along the southern edge of the site, which is an additional habitat for wildlife. Here the mix of species planted includes oak, birch, alder and hawthorn. These projects are due to be replicated right across the country and will eventually contribute significantly to restoration of Ireland's native forests.

Today in Northern Ireland the remnants of the great post-glacial forest cover just 0.04 per cent of the land. Special efforts are being made to protect these valuable reservoirs of biodiversity as continuously wooded areas hold special features, especially those associated with veteran trees and decaying wood. One such area is the Faughan Valley in the North Sperrin Hills in County Derry. Here there are narrow river valleys with corridors of wooded glens set in a landscape of farmland and flat-topped hills. The surviving ancient woods here are not in a pristine condition. Many were planted with exotic timber-producing trees or invasive species such as laurel to act as game cover for hunting. Increasing numbers of deer and lack of conservation management have led to other problems such as poor regeneration. A voluntary body, the Woodland Trust, which is active throughout Britain and Northern Ireland, has restored over 30 hectares of ancient woodland in the Faughan Valley with the help of volunteers and contractors and in partnership with local landowners. Over eight kilometres along the banks of the River Faughan and its tributaries have been planted with native trees in an effort to connect the fragments of ancient woodland. A further 61 hectares of land have recently been purchased by the Trust to increase woodland cover in the valley. The ultimate aim is to restore and connect the ancient woodland with a mosaic of different ages of native trees.[36]

The disappearance of Ireland's natural forest cover is closely linked with the loss of particular species which were dependent on this ecosystem. Although these trends are reflected across other groups of plants and animals it is easiest to focus on the best known and most intensively studied groups of Irish animals – birds and mammals. Of the 50 or so regular breeding birds known from forests, woodlands and scrub, most are relatively common, often occurring in hedgerows and gardens as well. At least three woodland species, that we know of, have become extinct in historic times. These were the capercaillie, a large turkey-like bird, and two birds of prey, the goshawk and the red kite, the latter having been recently reintroduced in the east of Ireland. On the plus side, three species of woodland birds are now recovering naturally from historic low populations although the reasons for these expanding numbers are only partly due to survival of their preferred habitats. These are the great spotted woodpecker, common buzzard and blackcap.

Among the larger forest animals introduced to Ireland in more recent historic times, the sika deer has become a major constraint on the recovery of native woodland. Originally from the north-eastern region of Asia, they were first introduced to Ireland in 1860 by Lord Powerscourt who maintained a small herd on his estate in North Wicklow. Here they bred and small groups were subsequently introduced to counties Kerry, Limerick, Fermanagh, Tyrone, Down and Monaghan.[37] In the early 1900s, as the great estates fell into decline, the deer escaped into the surrounding countryside where they multiplied and interbred with the closely-related native red deer. Today, these hybrid animals are widespread, especially in counties Wicklow, Kerry, Mayo and Tyrone. In the absence of large predators such as wolf and lynx, there are no controls on their numbers and the population continues to grow and expand. When present in large numbers, they browse the undergrowth in woodlands, eating the growing tips of young seedling trees and stripping the bark off larger saplings. The result can be a complete absence of regeneration to replace the mature trees as they die off. In certain parts of Ireland, it is nigh impossible to establish new native woodlands without deer fencing. Ciaran Fallon of Coillte Nature says,

> foresters I have spoken to on the continent strongly advocate organised and proactive deer hunting. In these countries open grassy areas (known as 'lawns') with high-chairs for shooting, are included in forest plans and hunting is actively promoted. I am convinced that we will not achieve our ambitious native woodland afforestation targets without addressing the deer challenge.

In 2009 the non-government organisation Woodlands of Ireland commissioned a comprehensive review of deer in woodlands as a basis for a proper strategy in dealing with the issue of deer damage. It concluded that the common species of deer in Ireland – red, fallow and sika – are increasing rapidly and in many areas are already at unsustainable levels. The report found that deer browsing is currently impacting significantly on both the economic and biodiversity values of forest habitats and predicted that these impacts will reach unsustainable levels if the current system of lack of management remains unchanged. These authors believe that the consequences of not addressing deer management are causing major problems for the conservation of existing native woodland, for the prospect of natural regeneration and for the necessary replanting of broadleaved trees as well as having many other economic impacts.[38]

Fifteen years later, foresters and forest owners still rely mainly on recreational hunters to try to reduce deer numbers. However, sport hunters tend to select mainly the larger stags, while reducing the number of breeding females is the only way that the population can be properly controlled. This will only be achieved with the establishment of well-funded and professionally staffed deer management units around the country. This is one of the proposals of a new Irish Deer Management Strategy Group which has also recommended establishing a National Deer Management Agency with a dedicated Programme Manager.[39] The recent proposal to classify sika deer as an invasive species would allow them to be shot at all times of year rather than just in the 'open season'.

There has never been a full census of deer in Ireland due to the difficulty of counting them in dense forests, but a new project at University College Dublin, under the title SMARTDEER, has produced some very good modelling of their distribution and density based on a combination of data such as culling returns and random sightings.[40] One recent innovation has been the use of heat-sensitive aerial photography using drones which can identify deer even in dense woodland or in darkness. This technique has been used to good effect in Baronscourt Estate in

County Tyrone and is informing the estate management about the real scale of deer control that is needed.

What are the prospects for restoring native woodland on a significant scale in Ireland? There are many bureaucratic hurdles and entrenched attitudes to overcome before this could become a reality. The ideal start would be to fence off the margins of lowland rivers and streams allowing natural regeneration of trees and shrubs. This would create long corridors of trees, linking existing patches of woodland and providing cover for woodland birds and mammals such as woodpeckers and pine martens. The tree roots would also have benefits for the waterways as permanent woodland can intercept fertilisers and other pollutants draining off neighbouring farmland. A similar approach in the uplands is to fence off gullies and river channels where there are already a few surviving trees. These provide the seed sources for natural regeneration of native trees and the fencing would prevent the saplings being grazed out by sheep or deer. Given the history of deforestation, and the dominance of farming landscapes, extensive return of large native forests seems unlikely in the short term but expanding corridors would create refuges for woodland species and a source of seed for a time in the future when the intensity of land use is reduced.

The magic of wandering at will through a mature native woodland in spring is something quite unknown to most Irish people today. However, it could become a new experience for future generations. As Paul McMahon wrote in his recent book *Island of Woods*, 'We need to develop a true forest culture.'

Reimagining the hills

When I first came to live in Wicklow over 40 years ago, I was fascinated by the mountains and I spent much of my spare time hill-walking. The Wicklow Mountains are the largest continuous range of uplands in Ireland and stretch for 60 kilometres from South Dublin into County Carlow. Although not high by international standards, the highest peak, Lugnaquilla at 925 metres, is Leinster's tallest mountain. In the 1990s, I was researching my first book, *Wild Wicklow,* and I was determined to explore all the higher parts and to find some of their most interesting wildlife. I trained in mountain leadership skills at an adventure centre near my home. I trekked from three different directions to the summit of Lugnaquilla, accompanied by the deep croaking call of ravens. But after a while, I stopped bringing binoculars as birds were scarce.

At that time, I thought that the mountains represented the most natural part of the east of Ireland. But something was missing. There was little woodland except for the uniform ranks of exotic conifers in state

forestry plantations. The only surviving native trees –
stunted rowans, birches and hollies – clung onto the steep
rocky slopes of gullies and cliffs where they were out of
reach of grazing animals. There were few large birds of
prey except for the occasional peregrine. Herds of deer
roamed the hills with no large predators to control them.
Groups of feral goats scaled the slopes of Glendalough
and neighbouring valleys, browsing any plants they could
find. But mostly it was the mountain sheep that kept the
vegetation short and quickly devoured any tree seedlings
that managed to establish. To encourage more grass and
support even more sheep, the mountain vegetation was
regularly burned and heather moorland was becoming
scarcer. As I got to know the mountains, I realised that
this was not the natural state at all. In fact, most upland
areas in Ireland were, and still are, severely overgrazed.
Without this pressure, the tree line should reach right to
the summits of some of the hills and mountains, limited
only in areas of unstable scree slopes. As long ago as
1945, the eminent ecologist Arthur Tansley recognised
that most of the upland vegetation in these islands 'owed
its present condition to the continual nibbling of sheep',
and this is equally true today.

Nevertheless, uplands form Ireland's largest expanses
of semi-natural landscape with approximately 29 per cent
of Ireland's landmass estimated to be over 150 metres
high and 19 per cent as unenclosed hills. Obviously, the
proportion of land is smaller the higher you go, with just
under six per cent of Ireland over 300 metres and less than

half of one per cent above 600 metres in altitude.[41] This
means that we have very little truly montane or alpine
habitat. Despite the virtual absence of native woodland
in the hills, they still support some habitats of high
nature value. For the purposes of the *National Survey of
Upland Habitats*, mountains were defined as unenclosed
areas of land over 150 metres and contiguous areas of
related habitat that extend below this altitude. However,
erosion, drainage, uncontrolled burning, overstocking
of sheep and extensive afforestation have resulted in the
widespread degradation of upland habitats, while wind
energy development and climate change are likely to
present further threats.[42] There are some 289 summits
in Ireland above 600 metres high, the majority of which
were sculpted by the glaciers which covered Ireland in the
last ice age. The gentler slopes are generally clothed in
either blanket bog or acid grassland which is maintained
by the ever-present grazing of sheep.

The highest peak in Ireland, Carrauntoohill, with
an altitude of 1,039 metres, lies in the centre of the
Magillacuddy Reeks. The stunning scenery all around
the peak marks one of our most spectacular mountain
ranges while to the south is the Black Valley which leads
down to the Killarney National Park with its surviving
native woodland. Although the Black Valley itself is
largely devoid of woodland today, it is full of placenames
containing the Irish word *doire*, meaning oakwood. Other
local names suggest that this valley once held a mixture
of oak, birch and holly woodland. Historical documents

from the 18th century show that these woods were being used for low-intensity grazing by cattle. However, the woodland-related placenames occur most frequently in cliffs, mountain fissures and very stony slopes, suggesting that the more accessible areas had already been cleared of trees by this time. The cause of this deforestation can be deduced from the widespread presence of charcoal production sites on the rocky slopes, dating to the period 1650–1800. Pollen records support the idea that deciduous trees were a significant presence in this landscape up to at least the mid-17th century and were mainly located on the rocky slopes up to 250 metres, forming a band between the wetter blanket bogs of the valley bottom and the heathland of the higher bogs. Dr Eugene Costello has advocated temporary fencing on these slopes to allow limited regeneration of woodland, particularly the rare habitat type sessile oak woods.[43]

The effects of centuries of sheep grazing in upland landscapes are also evident in the wild animal populations. Grouse are relatively rare in Ireland because the heather moorland habitat on which they depend is scarce and frequently burned. The mountain hare is present in such low densities that it does not provide sufficient food for large birds of prey such as golden eagles and the poison used in sheep-farming districts to control foxes also hits other non-target scavengers like buzzards. Ground-nesting birds of prey such as the merlin have disappeared from large areas of the hills, their nests often burned by the frequent spring fires or the eggs and young birds predated by foxes and crows that use the forestry plantations for cover. The merlins have switched to nesting inside the edges of conifer plantations but these nest sites can be lost quite suddenly when clear-felling takes place.

The effects of all the intense sheep grazing and burning have left a legacy of bare and eroded peat hags in places. Not only is this bad for biodiversity, but it also means that the exposed peat is emitting carbon to the atmosphere, adding to the already critical climate crisis. Faith Wilson is an ecologist who was engaged by an EU-funded project called SUAS to survey and report on the condition of a number of upland commonages in the Wicklow Mountains. Speaking at the final conference of the project in 2022, her assessment of the state of these areas was stark.

> Some upland areas have been repeatedly burnt and the result is bare peat that is haemorrhaging carbon and on which there is no grazing for livestock. The peat is being washed off the mountains with ridges eroding and a high risk of landslides. Burnt areas are invaded by bracken and every time the peat is burnt it gets harder and dryer. Reduced capacity to absorb rain means a higher risk of floods downstream and increased flood intensity in lowland rivers.

In the long term, the burning leads to colonisation by unpalatable plants like mat-grass and purple moor grass at the expense of heather and mosses. Faith says, 'These are managed landscapes that need proper management.

The legacy of upland burning in Wicklow and Dublin has been extremely damaging. The hills have been going wrong for a long time.'

In the Wicklow Mountains National Park, there are a number of ongoing peatland restoration projects to address this. With Hugh McLindon of the NPWS, I went to visit one of these sites on Barnacullian, just north of Lough Ouler, where the peat is in the throes of total collapse. It is completely de-vegetated and is just washing and blowing away, causing major issues downstream. In order to arrest this, National Park staff have fenced off areas to reduce the grazing pressure. The physical impact of animals trampling and rubbing off the bare peat causes damage as well as preventing new growth taking hold. The mobility of the bare peat is phenomenal. Hugh says, 'it's like the Sahara at times up there with peat moving around like sand dunes in the desert'. Following the model of Moors for the Future, which has been tackling similar problems in the Peak District of central England, they cut heather brash on neighbouring hills, using a double chop silage harvester, for spreading on the bare peat at Barnacullian. This has the added benefit of achieving suitable vegetation management on the dry heath at the donor sites while providing material for stabilising the bare peat on the restoration sites. The brash is excellent at creating a mat that prevents peat mobility and allows for mosses and grasses to establish. In places they spread lime, fertiliser and grass seed to create nurse or sacrificial grass and also to stabilise the bare peat. This grass then dies away as the liming is stopped and the acidity increases again towards a natural level in the peat. After about three years, the native mosses and grasses are already establishing and will have nearly completely colonised the site. By the fifth year, the site should be fully revegetated with native species.

As it is the power of the water running off the site that does a lot of the erosive damage, the key to success is to slow this down. Retaining the water for longer within the bog allows *Sphagnum* mosses to reestablish themselves. Without *Sphagnum* the bog has no chance of becoming 'active' and sequestering carbon again. In the largely flat raised bogs of the midlands it is easier to raise the water table to kick-start this process, but in blanket bogs the gullying and slopes make it challenging to lift the water levels back to the surface uniformly. Effectively, it becomes a mosaic with different solutions being applied across the site. There are places where timber dams are being used, while in other channels the teams have constructed stone dams.

The first season of restoration work was carried out on Barnacullian in 2022 and this has begun the process of re-vegetating some three hectares. In the same area, approximately 200 timber dams and about 30 stone dams have been built to allow for rewetting across another four hectares. A second phase of the project started in 2023 and this is expected to revegetate three hectares and rewet another four hectares further up the ridge. An innovative use of recycled local sheep wool, virtually worthless in the marketplace, is being used here to create biodegradable dams in the channels. Permanent monitoring stations (quadrats and photo posts) have been put in place, and the return of the vegetation will be monitored from year to year. Hugh says, 'The team is constantly building knowledge on the local conditions and how different techniques work or not, but it is still early days. There are over 300 hectares of bare peat in the Wicklow Mountains National Park alone, so we'll be at it for a while.'

On the nearby Liffey Head Bog, a much larger project covering 100 hectares is well underway. Here peat dams, plastic dams, coir logs (a coconut byproduct) and leaky timber dams are being applied as conditions allow to stem the loss of peat. The National Park staff and contractors have been supported by teams of volunteers. The peat surface here is so damaged that there is a high risk of landslides when using machinery so much of the restoration work has to be done by hand. The work is being carefully monitored with instrumentation in the peat measuring water levels as well as the release of greenhouse gases. Water quality is especially important here as this area is part of the catchment for one of Dublin's main water supplies. Once the results of the monitoring are available it is planned to significantly expand the scale of the work.

A similar approach to restoring upland bogs has been used in the Cuilagh Mountains, shared between counties Fermanagh and Cavan, through CANN (Collaborative Action for the Natura Network), a large EU-funded project in Ireland, Northern Ireland and Scotland. The project

team worked on a couple of areas only a short distance
from a boardwalk on the Fermanagh side of the border.
Here there are significant areas of blanket bog erosion
which had left deep gullies, peat hags and expanses of
bare peat. Simon Gray of Ulster Wildlife told me it was
possible that this was due to historical overgrazing or the
repeated passage of stock but the reasons behind it were
not clear because they could have ceased a long time ago.

The team worked with the local landowners here,
training them in restoration techniques and overseeing the
works over the winter of 2020–21. They did a combination
of peat hag reprofiling and using coir logs to slow the flow of
water through eroding areas. They tried a patch of heather
brash on an area of bare peat to kick-start revegetation
there too. They had to helicopter in most of the materials
and guide two diggers well up into the high parts of the
mountain, which made for some difficult work. The results
there have been promising so far with bog cotton recovering
well in previously bare areas and *Sphagnum* moss even
starting to colonise the coir logs themselves after 12 to
18 months. They need to spread a lot more heather brash
to cover areas of bare peat that will never become fully
rewetted. The other aspect of the project on Cuilcagh was
working with a local farmer to block a number of drainage
channels on his area of blanket bog high up the hill. This
was a much larger area of about 80 hectares over which the
drainage had taken place. The drains dry out the blanket
bog in these areas, particularly in the summer periods.
They trained the farmer in drain-blocking techniques and

worked with him to install over 300 dams across the area in 2020–21. The involvement of local farmers in this work is a good sign for the future as their community will have responsibility for it in the long term.

Percy French immortalised one mountain range in the north-east of Ireland in his famous lyrics, 'where the Mountains of Mourne sweep down to the sea'. Today the Mourne Mountains in County Down lie bare and largely deserted. Where once there were native forests with wolves and wild boar, eagles and owls, today there is unending sheep pasture stretching as far as the eye can see among the classic hills and valleys. The Silent Valley, which holds Northern Ireland's largest reservoir, has some woodland in its lower part but this is an exception. The 12 peaks in the Mournes have been designated as an Area of Outstanding Natural Beauty and the National Trust has joined with others to manage a large section of the mountains. There have already been some practical projects such as restoring woodland and upland heath in Annalong Valley and path repair along the Glen River. However, these are hard to find among the purple moor grass and heavily grazed heather. Shifting baseline syndrome has led us to believe that this is the normal situation. During our lifetimes these areas seem to be unchanged so we assume that it is a natural landscape.

A key aspect of looking after the Mourne landscape is the heathland management project. Increased recreational use, grazing and wildfires have caused significant damage. Consequently, these combined pressures have altered the

species composition and the condition of the habitat.
The Healthy Heathland project has involved refining best
practice approaches to suit the Mournes. Eroded peat
hags, for example, are an ever-increasing problem. The
creation of lots of dams helps to improve water retention
and promote regeneration, thus addressing this problem.
Both stones and large coir rolls are used to create dams in
the high Mournes. In 2021, a large fire damaged almost
300 hectares of land in the Mourne Mountains, including
part of Slieve Donard, Ulster's highest peak at 849 metres,
and an area once rich in flora and fauna. Vegetation was
destroyed and species diversity reduced. Since then, rangers
have been trialling different methods to rejuvenate the
land and bring it back to full health for the plants and
animals that live there. Now, working in partnership with
a tenant farmer, a herd of Luing cattle has been added
to the restoration efforts. The herd of six cows tramples
bracken and reduces the area of purple moor grass that
has spread since the fire, providing the space for heather
to return, and creating habitats for newts, lizards, ground-
nesting birds and hares. The cattle wear special collars with
GPS tracking which allows 'virtual fences' to be created.
This means grazing can be targeted at particular areas of
the mountains without the need for intrusive fences, while
maintaining high levels of animal welfare.

Similar 'no fence' technology has been used in a project
on Howth Head, within sight of the capital city of Dublin,
where gorse scrub and bracken have become extensive
due to a history of intensive fires, reducing the area of

heather and other more diverse vegetation. In 2021, gorse fires on the headland burned an area of approximately 26 hectares over several weeks, despite efforts to contain them by Dublin Fire Brigade and the Irish Air Corps. The publicly owned land on the headland includes the summit and several ridges as well as large areas of cliff and coastal heathland. The local authority here, Fingal County Council, decided that the best and most ecologically sensitive way to control the spread of gorse was to reintroduce a herd of goats. These domestic animals are very effective browsers of even the tough stems of gorse and, by using GPS collars, they can be focussed on particular problem areas without the need for fences. Twenty-five Old Irish Goats, a rare and localised breed, have been relocated from the hills of Mulranny in County Mayo to Howth Head, where they are being used to reduce gorse cover in an area that has been plagued by wildfires. Melissa Jeuken, who has previously tended goats on her family's farm in the Burren, County Clare, is employed to manage the herd and has attracted considerable publicity for this innovative project. The Council's Biodiversity Officer, Hans Visser, said, 'We have divided the landscape into smaller pieces, by means of fire breaks. But those fire breaks do need maintenance as well. We can use machinery to do this repeatedly – or the goats can graze those fire breaks, which is much cheaper and much more sustainable.'

On an autumn day, I joined a walking group to visit the spectacular valley of Luggala in the centre of the Wicklow Mountains. The valley has two classic glacial lakes surrounded by high cliffs, upland blanket bog, dry heath, acid grassland, meadows, conifer plantations and mixed woodlands, some of which are dominated by native oak. For over a century, this 2000-hectare landscape was owned by the Guinness family, but it was purchased in 2019 by Luggala Estate Ltd and now has three full-time staff who are undertaking a large restoration project. Reaching the farm buildings at the bottom of the valley we met Michael Keegan, the farm manager, who explained the project to us. Before the work started, a baseline study of habitat conditions informed an environmental management plan for the entire estate. Actions undertaken to date include heather cutting on the higher ground to benefit red grouse and other species, treatment of the invasive rhododendron plants in the woodlands, bruising and treatment of bracken on the slopes, with repair and restoration of the traditional drystone walls on the estate. The staff have gathered seeds from native trees and shrubs for propagation followed by woodland restoration and tree-planting in gullies around the rivers. This necessitates deer fencing and management of the large sika deer population in the valley including the culling of unhealthy animals. Michael says, 'In terms of reforestation, we are targeting areas of the estate where there would have been woodland previously and we are replanting in those areas. The idea is to join the areas of woodland creating wildlife corridors throughout the estate.' Future plans include the incorporation of cattle to the land as a means of conservation grazing, rewetting of blanket bog areas, additional planting of native trees and conversion of plantations to broadleaves. Michael's own attitudes to upland farming have also been transformed over the past two years, resulting in changes to farming practices on his home farm as well as at Luggala.

The Scottish mountains have a very different land ownership structure to those in Ireland. Much of the Highlands and Islands are divided among a few large landowners, many of whom are not even resident in the area. Between 1750 and 1860, the great Highland Clearances removed a large proportion of the native population with forced evictions of a significant number of tenants. The Highland Potato Famine struck towards the end of this period causing widespread starvation and destitution. Emigration of many was assisted by the landowners, who paid the fares for their tenants to be shipped off to America. The small farms were replaced by extensive sheep pastures and deer numbers increased as the great estates were making money from deer stalking and grouse shooting. All of this had massive impacts on the Scottish pinewoods, an ecosystem that is unique in Europe. While small patches of ancient trees remained in some of the less accessible locations, the nibbling mouths of sheep and deer ensured that there was little or no tree regeneration.

Now, two centuries later, a large swathe of the Scottish Highlands, stretching from the west coast to Loch Ness in the east, is to be managed as part of a 30-year project to restore nature. The Affric Highlands initiative aims to increase connected habitats and species diversity over an area of 200,000 hectares incorporating Kintail mountain range and glens Cannich, Moriston and Shiel. Plans include planting trees, enhancing river corridors, restoring peat bogs and creating nature-friendly farming practices. The project has been launched after two years of consultations and discussions between local communities and conservationists from the charity Trees for Life. It is a community-led effort to restore nature over a large area, which the organisers hope will be a catalyst for social and economic regeneration. Native wildlife set to benefit includes a range of river species such as salmon, trout, osprey and otter, as well as montane species such as golden eagle, short-eared owl, mountain hare and red grouse.

The red grouse is one of the most characteristic bird species of upland habitats, where it remains throughout the year. It has traditionally been a quarry species and large parts of the Scottish mountains are intensively managed to boost the populations for shooting. In Ireland, as elsewhere, the grouse population is heavily dependent on the condition and variation in structure of ling heather. Some 60 per cent of red grouse in Ireland are found on upland commonages. This involves multiple landowners sharing the grazing over a mountain area that is unfenced. In 2021–22, the National Parks and Wildlife Service commissioned a survey to determine the national population (and any change since 2006–08), including threats and pressures on red grouse nationally. While results were still being finalised in 2023, the population has not recovered, with a further 14 per cent reduction in range at ten-kilometre square level since 2006–08.[44]

Over the last decade, there has been a more concerted effort to reverse the decline of red grouse populations in some parts of Northern Ireland and the Republic of Ireland. In 2007, the Glenfarne Gun Club identified Boleybrack Mountain, in the south of County Leitrim, as an area that should be managed in order to protect and enhance its important grouse population. Firstly, the project reduced the pressures on the grouse due to random burning, fox and crow predation and human disturbance. The bigger challenge was to find a sustainable management pattern that incorporated the needs of the farming shareholders and greater biodiversity. The gentle slopes of the plateau are dominated by mountain blanket bog with wet heath, while drier areas on steep ground and above the cliffs are dominated by dry heath. There are also areas of purple moor grass meadow as well as poor fen and flush. In the winter it is pretty wet and not easy to walk over. In general, red grouse require tall heather for shelter or nesting and prefer shorter heather for feeding.

Heather management by burning is carried out here in a controlled manner each year. No burning takes place on blanket bog habitat that is still actively growing with *Sphagnum* mosses. A small percentage of dry heath and some degraded wet heath is burned on a long-term cyclical basis. The project team initially tried to control and manage the heather by manual strimming but this proved to be too labour-intensive and was not an option on this terrain. A mosaic of small clumps of scrub and other vegetation is not burned on the dry heath areas in order to maximise the botanical variety available and provide an essential source of food for grouse during the 'feeding bottlenecks' and any prolonged period of snow cover which may be encountered every decade or so.

In the past, gamekeepers were employed in some mountain areas to reduce predation on upland birds such as grouse by killing all predators such as foxes, stoats and birds of prey. As this type of systematic culling is rare in Ireland today, some small predators such as fox and hooded crow have 'unnaturally high' population levels, as they have benefitted from the extermination of wolves and golden eagles, former apex predators in Ireland. The smaller predators have also prospered in the hills due to the increased cover provided by widespread forest plantations. Ground-nesting birds such as grouse, curlew and golden plover are among the native wildlife species that have suffered from this increased predation.

Entering the valley of Glenveagh in the heart of Donegal I was reminded of some classic views of the Scottish Highlands. I made my first visit to Glenveagh, many years ago, walking in from the hills to the west. Approaching slowly down the valley, the ancient woods opened out to reveal a deep lake, stained dark brown by peat washed down from the hills. Just like in the Scottish Highlands, the mountain vegetation here is grazed by herds of red deer. By 1845, the native herds had been hunted to extinction but later in the nineteenth century, various introductions from Scottish and English stock and from Irish collections re-established the red deer on these lands. By 1897 a deer fence had been completed to keep the herds from straying so that deer stalking and hunting could be carried on by the owner and his wealthy visitors. During the 1950s and 1960s, when detailed records were kept, 20–40 stags (males) and 10–30 hinds (females) were shot each season. However, the absence of natural predators such as wolf and golden eagle, due to persecution, allowed the deer herds to grow and spread beyond the estate. The combination of wild deer and growing numbers of mountain sheep caused serious overgrazing across the mountain ranges. By the 1990s it was considered that a herd of 450 deer was 'optimal' for the Park and, to maintain this population level, a proportion of the herd was culled annually by the NPWS.

By 2017, when the habitats in the Glenveagh and Derryveagh mountain areas were assessed, all the vegetation types were considered to be in 'unfavourable condition', using standard EU terminology. This included all the blanket bogs, wet heaths, dry heaths, alpine and boreal heaths as well as various rare types of mountain vegetation. Achieving the EU standard of 'favourable conservation status' is the overall objective to be reached for these habitat types, which are listed in the EU Habitats Directive. This means that the habitats must be prospering and have good prospects of continuing to do so. How this is to be achieved is not so clear. Dr Ciaran O'Keeffe was Park Superintendent here for 15 years in the 1980s and 1990s when I first visited Glenveagh National Park. He later became a Director of the NPWS but is now enjoying his retirement. He told me that, in his time at Glenveagh, the main conservation issues here were rhododendron infestation in the park woodlands and overgrazing by large numbers of sheep and deer in the mountains. The red deer at Glenveagh were kept in the estate by a high deer fence but this was frequently breached over the years by snowdrifts, broken wires and undermining by various animals.

The deer are on the hills for much of the year but in winter many come down to shelter and graze in the native woodlands of the park. The deer population reached a peak in the 1990s so the NPWS adopted a policy of culling to reduce the pressure, but deer outside the Park were increasing and they supplemented the population inside. The deer population here was reduced from approximately 1,000 in the 1980s to around 360 in 2017. Rather than cull the deer herd to a very small number, several fencing exclosures were established in the woodlands near Glenveagh lake to prevent deer from restricting woodland regeneration. Monitoring inside these fenced areas found a succession of regenerating trees including birch and holly.

The mountain bogs of Glenveagh National Park have not been entirely free of other damage either. The ecologist Pádraic Fogarty related in his landmark book, *Whittled Away*, how he was shown four areas within the northern boundary of the National Park where the vegetation had been stripped away over hundreds of square metres leaving bare earth and muddy pools.

> The telltale parallel lines of extruded peat could only be left in the wake of the sausage [turf-cutting] machines that can devastate a bog in a matter of hours. In one location the destruction was embellished with a liberal sprinkling of waste tyres and other debris – as if to emphasise how this place is seen as little better than a landfill site.[45]

Writing to the government to present his first-hand evidence of the damage, Fogarty was told only that there were traditional turbary rights (small-scale turf-cutting with hand tools) in certain parts of the Park. Is it acceptable that damaging activities such as commercial peat cutting, burning and intensive sheep grazing are allowed to continue in this and other National Parks, which are supposed to have the highest national value for nature conservation? These pressures present a major challenge to bringing these habitats into better condition.

Back in those optimistic years at the start of the new millennium, an exciting project was launched at Glenveagh to reestablish a viable golden eagle breeding population in the northwest of Ireland. The project, managed by the Golden Eagle Trust, was originally part-funded by the Irish Government as part of its National Millennium Celebration. Scotland was chosen to supply the birds, under licence, as this was the nearest eagle population that was likely to be genetically similar to the original Irish birds. In those early years, I travelled to Donegal with a friend during a wet summer when the rain never seemed to stop. We had arranged to meet Lorcan O'Toole, Project Manager with the Trust. Lorcan gave us some tips on where to find the eagles as his radio-tracking equipment indicated an approximate location. So, we headed off into the Derryveagh Mountains with the mist hanging low over the summits. After several hours walking we finally found a young eagle that had been sheltering from the rain in a stunted oak tree, and it soared away over the hills. These were the early days of an ambitious project to reintroduce these magnificent birds of prey to the Irish landscape. Nearly a century earlier they had been eliminated from our skies by a combination of shooting, poisoning and egg-collecting at a time when any wild predator was seen as a threat to farming.

Bird of prey reintroductions are based on the knowledge that they generally try to breed close to their own original nest sites, and consequently reintroduced birds are usually released in protected areas where they can subsequently nest safely. In conjunction with the Scottish Raptor Study Groups, the Golden Eagle Trust had planned to remove up to 12 wild partly grown chicks each year from nests in Scotland when they were five to six weeks old. At this stage, the eaglets can feed themselves and keep a stable body temperature. They were quickly transported to Donegal and placed in specially designed avian cages. The cages contained artificial nest platforms and perches. The birds were fed dead rabbits and crows through a secure hatch and sleeve so that no human contact took place. It was essential to avoid human imprinting at all costs. Once released into the wild, the newly fledged eagles were provided with game in food dumps for several months as their parents would do in a natural population. The objective was to release enough birds to ensure a sufficient number of individuals survived to breeding age and formed a viable population.

But the project ran into a number of difficulties right from the start. The Scottish authorities were worried about some high-profile poisoning incidents in Ireland and there was increasing concern about the status of the donor Scottish golden eagle population. Poor breeding success there meant that there were insufficient nests with the required number of chicks to allow collection of the target numbers and only 66 young eaglets were collected from Scotland over the period 2001 to 2012. The question was whether this small number of pioneers would be sufficient to establish a self-supporting population in Ireland. By 2007, one pair had fledged a single chick in Glenveagh National Park. This was the first successful breeding of golden eagles in the Republic of Ireland in approximately 100 years and in the following years a succession of chicks flew from wild nests. However, by 2011, only five pairs had managed to produce eggs at least once.

Illegal poisoning of birds of prey remained a problem, but an equally worrying factor was the poor quality of the habitat in the uplands of Donegal, and it does not appear that agricultural policies or conservation designations are improving this situation. Survey work in 1990–91, before the reintroduction project began, suggested that there was indeed sufficient live prey to support eagles in Donegal. However, recent evidence collected by Fiona McAuliffe, a student at Queen's University Belfast in 2017–18, showed that the current densities of the preferred prey, such as mountain hare and red grouse, are only about one-third of the necessary levels to support a viable population of golden eagles. The paucity or absence of these prey species in some Donegal uplands can be taken as an indicator of poor habitat quality which has resulted from intensive grazing by sheep and deer as well as uncontrolled burning. The only way to improve this would be to reduce the grazing pressure, allowing vegetation and the dependent prey species to recover.[46]

Most ecologists agree that the uplands of Ireland are in very poor condition when their biodiversity is considered. Centuries of intensive overgrazing and burning have left the habitats so modified that very little remains of the natural vegetation. Even heather moors probably spread as a result of the clearance of scrub and trees thousands of years ago. The pressure continues today with most of the higher land owned in common by multiple landowners and grazed by free-ranging flocks of sheep. In some places – notably Wicklow, Kerry and Donegal – there is the extra problem of unmanaged deer herds adding to the grazing pressure. Added to the difficulties faced by large birds of prey is the continued use of poisons which are mainly connected with sheep-farming. These are long-term difficulties that will take decades to change and any attempts at restoration will have to accept that large predators need extensive home ranges and cannot be contained within small, protected areas.

Nevertheless, the National Parks in the mountains of Kerry, Connemara, Mayo, Clare, Donegal and Wicklow are probably the best places to test out new ways of restoring upland habitats and species. Glenveagh National Park is one where a number of conservation issues – overgrazing, unauthorised turf-cutting and invasive species – are being tackled and lessons have been learnt here. Innovative peatland restoration work in the Wicklow Mountains National Park is testing methods that could be used much more widely in the uplands. Intensive efforts have been made to remove the invasive rhododendron from the important woodlands of Killarney National Park, with limited success. The national parks also offer great opportunities to educate hundreds of thousands of visitors about nature restoration and to seek their support. If some of the money that the National Parks generate from tourism could be diverted into restoration work on the degraded habitats this would be a welcome move. Removing the extensive plantations of non-native conifers from the National Parks and restoring native woodland would also help to diversify the habitats. Any habitat restoration needs to be done on a landscape scale rather than in a piecemeal fashion but a key component will be changing the attitudes of the landowners who have grazed sheep in the mountains for centuries.

Our mountains are part of Irish culture as well. Saint Kevin, who was the abbot of the famous Glendalough Monastery in the Wicklow Mountains, is reputed to have held a blackbird in his hand for the entire breeding season while she laid eggs and raised her young. Mountains also became a refuge for Irish rebels in times of strife such as after the 1798 rebellion. The 'holy mountain' of Croagh Patrick has, for generations, been a place of pilgrimage and prayer. The mountains provide space for recreation as well as quiet places far from the bustle of everyday life. These cultural values, as well as their importance for nature, make it vital that nature restoration is given a high priority in our uplands. The pioneering American naturalist John Muir wrote in 1876, 'Thousands of tired, nerve-shaken, over-civilised people are beginning to find out that going to the mountains is going home; that wildness is a necessity.'

Farming with nature

One of the abiding memories from childhood is of my father mowing the meadow around our house in the summer using the old hand scythe that still hangs in my workshop. The swishing of the scythe, little changed since the Iron Age, cut through a sward with buttercups, cowslips and orchids as well as a wide range of meadow grasses. The sound of the stone sharpening the blade and the sweet smell of new-mown hay are still vivid memories. For days afterwards, the family would be drafted in to toss the hay, using the wind to dry the grasses and building the results into small haycocks to be used as winter fodder for our cows. This was in the days before the widespread use of chemical fertilisers when our meadow received only animal manures to maintain its fertility. As a result, it was full of wildflowers and alive with butterflies and bees. I remember finding a hedgehog in the long grass, foraging on slugs and worms.

The Emerald Isle is a romantic name for Ireland, often used by people in other countries. Although rooted in the past, it certainly reflects the aerial view of the landscape if you arrive by plane on a clear sunny day when the patchwork of green fields, hedgerows and forestry is laid out below as far as the eye can see. Ireland today is predominantly a country of grass. Some four-fifths of lowland farmland is managed as grassland and the mountains are mainly covered in a mosaic of acid grassland and blanket bog. In addition to the economically important grasslands, there are also several grassland types of high nature value including orchid-rich grasslands, species-rich neutral grasslands and wet grasslands such as the Shannon Callows. Even the sand dunes and clifftops around the coast are dominated by grasses.

A century ago, species-rich grasslands were the typical kind of farmland in Ireland but, unfortunately, these have become a rarity in the Irish countryside. A survey of semi-natural grasslands, undertaken for the National Parks and Wildlife Service between 2007 and 2012, resulted in the mapping of 1,192 grassland sites covering over 23,000 hectares.[47] Over half of this was classified as wet grassland, most prevalent in the west of Ireland. However, many of Ireland's natural grasslands have been highly modified with the use of chemical fertilisers, reseeding with alien species of grass or such heavy grazing that few wild plants or animals can survive in them. About 28 per cent of hay meadows and 31 per cent of calcareous grasslands, such as those in the Burren, have been lost in the previous six years, according to the NPWS. The skylark, a characteristic bird of unimproved grasslands, has declined dramatically and is now on the list of Birds of Conservation Concern in Ireland. Corncrakes and breeding waders of grasslands, such as the curlew and lapwing, have also suffered catastrophic declines according to Birdwatch Ireland. So too have many species of grassland insects – butterflies, moths, bees and hoverflies – as reported by the National Biodiversity Data Centre.

Grazing is essential to maintain open grasslands. Abandoned grasslands quickly become invaded by scrub and trees. The wonderfully diverse limestone grasslands in parts of western Ireland are maintained this way by the munching of herbivores. Grazing is also important for some animal species that depend on open vegetation. Many butterflies, bees and other pollinators depend on the nectar produced by grassland flowers. Choughs are more common in Ireland than in any other West European country, partly due to the prevalence of short clifftop grassland which provides them with the ideal ground conditions to probe for their insect food. By contrast, the corncrake, a classic bird of traditional hay meadows, prefers to nest and feed in longer grass through which it can move, barely rustling the stems.

Important species-rich grasslands like those in the Burren need carefully managed grazing if they are to maintain their biodiversity value. The Burren occupies an unusual rocky landscape of north Clare and south Galway that is best known for its bare limestone pavements, but it contains much more. The major habitats in the Burren also include orchid-rich calcareous grasslands, limestone heaths, scrub and woodlands, wet grasslands, turloughs, calcareous springs and fens. Here it is common to find a diverse mix of plants among the limestone outcrops, some that are typical of arctic habitats and others from the Mediterranean. Most famous is the deep blue colour of the spring gentian which has been adopted as a symbol of the Burren in many products and tourist enterprises. The special value of the Burren is the survival of so many relatively rare habitats over so large an area, offering excellent 'connectivity' in contrast with the fragmented nature of such habitats elsewhere.

One of the keys to management of the Burren is the long-established practice of transhumance – the ancient tradition of moving livestock from one grazing ground to another in a seasonal cycle. Frosts are rare due to the warming effects of the Atlantic Ocean and growth continues in the shorter days, making the hill pastures available for grazing throughout the winter. Cattle graze on the native grasses while most of the wildflowers are dormant over winter, preventing them from being damaged. Winter grazing prevents the spread of scrub thus keeping the diverse limestone grasslands open and rich in plant and animal species. One autumn, I joined in the Winterage Festival which is held here each year, led by the local farming community, to celebrate the ancient practice of moving livestock to the Burren Hills. In typical West of Ireland rain, a large crowd gathered on the farm in the lowlands for a welcome cup of tea. Then we marched in communal procession after the cattle herd as the farmer and his many followers drove the animals to the uplands for the winter.

This farm is part of the pioneering Burren Programme, formerly the Burren Farming for Conservation Programme. Unlike many agri-environment schemes, it is farmer-led. Landowners nominate and co-fund conservation actions on their own farms and are generally free to manage the land as they see fit. The Burren Programme is results-based so that those farmers who deliver the highest environmental benefits are rewarded most by the scheme. Conservation becomes as much a product for the farmer as the livestock produced. Farmers are given the freedom to deliver the required outputs using their own skills, experiences and resources, as best fits their own farms and circumstances. This flexibility means that the Burren Programme is capable of dealing with most situations which can arise from farm to farm and from year to year.

One offshoot of the Burren Programme has been the setting up of the Burrenbeo Trust, a voluntary grouping of farmers, tourism operators, naturalists and other local residents to create greater appreciation for the Burren landscape, all based around a model of community-led conservation. An initiative of Burrenbeo is the Farming for Nature network which links together like-minded farmers throughout the country who are practicing good nature restoration management on their farms. This is a non-profit initiative whose mission is to support, encourage and inspire landowners who farm, or who wish to farm, in a way that will improve the natural health of our countryside. There are over 70 Farming for Nature ambassadors throughout the country now. Among them are Cathal and Bronagh O'Rourke who manage a 500-acre farm in the Burren. The farm is a mix of intensive grassland, species-rich grassland, mature hazel woodland, limestone pavement, and turlough. Part of the farm lies on the shores of Lough Bunny and it is within the Burren National Park. Here the farming practices help to support nature on the land. These include clearing scrubland to link up grassland areas, grazing cattle on the winterage to encourage the growth of native Burren flora and reducing inputs in the improved agricultural areas of the farm. Realising the potential of this farm for education, Cathal and Bronagh have set up a project called the Burren Farm Experience which runs food trails, guided walks and group tours. They describe their way of farming as a lifestyle which respects and enhances the land of which they are custodians.

During the 1970s, when I was working in Northern Ireland, I often visited the limestone country of County Fermanagh in the westernmost part of the province, to explore some of the rich wildlife habitats in summer. I was especially impressed by the flower-rich meadows that surrounded Lough Erne at that time. Lying in the grass to photograph the flowers I was struck by the hum of life as countless insects and other invertebrates were active in the vegetation. Five decades later many of these wonderful habitats have disappeared due to modernisation of farms. The Save our Magnificent Meadows project was set up to protect and restore species-rich grassland across the UK and includes County Fermanagh. It was initially led by the voluntary organisation Ulster Wildlife from 2014 to 2017, but since then it has been delivered with support from Butterfly Conservation, the local district councils and the Northern Ireland government. Hay meadows, once a feature of every parish in Northern Ireland, are now an increasingly rare sight. Over three million hectares of wildflower meadows have been lost across the UK so far and they are still being destroyed. Despite the loss of the vast majority of these meadow habitats, a significant proportion of what remains in Northern Ireland is found in Fermanagh and west Tyrone.

One of the project's success stories is the restoration of Navar Meadows in Fermanagh. This four-hectare lakeshore site, owned by the Forest Service, includes five small meadows. In 1960 this was described as the most species-rich grassland in Fermanagh, but over the last decade, the lack of suitable management led to the development of a thick layer of thatch, rush infestation and a complete loss in species diversity. A conservation action plan was produced for the site and a local farmer, George Ferguson, was approached to take a four-year agricultural lease on the fields. The meadows were cut, branches were cleared from gateways and works were completed to improve access. Three years later, through a combination of yellow rattle seed reintroduction and suitable agricultural management, a survey showed the restoration was a success, with an increase in positive indicator species such as common spotted orchid, meadowsweet, bugle and meadow vetchling. Navar

Meadows is now managed as a hay meadow, but the challenge will be to sustain this conservation management in the longer term.

At a different scale, Castletown House in County Kildare is one of the finest surviving classical mansions in Ireland. After a period of decline in the 20th century the house was taken over by the Irish government in 1994 and the entire estate is now managed by the OPW. On a warm summer day, I was shown around the habitats in the demesne by Rory Finnegan, the Head Gardener, who is also an experienced naturalist. The OPW has been managing over 32 hectares of former agricultural grassland here as a summer meadow since 2007. This is cut once a year in September and all vegetation is bailed and removed. This reduces the fertility which encourages native wildflowers and grasses. No fertilisers or pesticides are used, as these would negatively impact the biodiversity of the grasslands. Mown pathways wind through these meadows, allowing the public to appreciate the flora and fauna at close quarters. Rory pointed out with pride some orchids among the species-rich grassland. He explained that the introduction of yellow rattle has had a beneficial effect as it thins out the grasses allowing other plant species to colonise. The grassland management has yielded excellent results over the last ten years. The plant communities have become more diverse and there is a very marked increase in the populations of yellow rattle, pyramidal orchid, lady's bedstraw and cuckoo flower.

At Castletown, dandelions carpet the ground in spring, an important food for bees emerging from hibernation. A small population of devil's bit scabious grows on south-facing banks near the River Liffey. These plants flower late, in September–October, and are brilliant late sources of nectar for the all-important pollinator insects. Winning the Green Flag for Parks Pollinator was, for Rory and the staff here, a validation of ten years of work managing the demesne in an ecologically sensitive way. The erection of the 'Managed for Wildlife' signs, provided by the All-Ireland Pollinator Plan, received a very positive reaction from the general public. Mature parkland trees that have died have been left in the meadows to decay naturally. Holes of different diameters are drilled in these trees to provide nest sites for solitary bees. On a south-facing bank, vegetation has been scraped away to provide another opportunity for these little known but important pollinators. Rory showed me a wild honeybee nest in a hollow veteran tree on the riverbank. We could hear the buzz of the colony within the tree. A large population of the red mason bee have made their nests in a west-facing wall of the house itself.

Managing the estate for biodiversity has been a very rewarding experience for Rory and the other staff of Castletown Demesne. He recalls that, at the height of its glory in the 19th century, there were over 200 ground staff working on the demesne. All the work was done with hand tools and horse-drawn machinery. Rory says,

We should acknowledge and identify with our ancestors who depended on wildlife and natural resources. Today we benefit from these for recreation and amenity but we have a duty of care to all the wildlife which depends on these same resources for their very survival into the future.

Today, under Rory's guidance, the extensive lands of Castletown Estate have found a new purpose, forming an island of diverse wildlife habitat close to the capital city.

In the rich farmland of Munster, the BRIDE Project is an innovative scheme that has been designed to conserve, enhance and restore biodiversity in lowland intensive farms in the River Bride valley of east Cork and west Waterford. BRIDE is also an acronym for 'Biodiversity Regeneration in a Dairying Environment'. The project provides participating farmers with specifically designed farm habitat plans that identify the most appropriate and effective biodiversity management options for their farms. This project uses a results-based approach (in contrast to traditional agri-environment schemes) and offers extra payments for increased biodiversity gains. This gives a signal to farmers to encourage hedgerows to grow with side trimming being the optimum management choice. Grassland with more plant species gets a higher payment than grassland with fewer plant species. A higher price for a better-quality product is a very familiar concept to farmers and the BRIDE Project applies this principle to the management and improvement of habitats. This also means that farmers are paid for the ongoing management of existing habitats and this is an important feature of the project.

Donal Sheehan, Project Manager and dairy farmer, notes that the BRIDE Project was 'designed by local farmers for local farmers and this is one of the most important distinguishing features of the project'. The BRIDE Project is open to all landowners within the Bride valley and although dairy farmers constitute the largest group, participating farmers also come from the beef, equine, sheep and arable sectors. The Project has been operating since 2018 and has enhanced biodiversity on 43 intensively managed farms by introducing a range of innovative measures that do not impact negatively on farm productivity. Particular attention is focussed on declining farmland species such as barn owl, stock dove, skylark and yellowhammer, and new wildflower mixes are being formulated to improve foraging habitats for threatened pollinating insects such as bees, butterflies and hoverflies.

Farmers can choose from a range of options and, in this way, they play an active role in the design of their own habitat management plans. The measures (including pond creation, skylark plots, pollinator plots and targeted nest-box schemes) are specifically designed to maximise biodiversity on each participating farm. New ponds have been created on 37 farms, new native woodlands on 31 farms with six kilometres of new hedgerow planted. Hedgerow management measures encourage reduced trimming and instead, farmers are encouraged to let hedgerows grow and mature, thus providing more flowers for pollinating insects and, in turn, more berries and seeds for birds and small mammals. Two-metre-wide field margins are not sprayed or fertilised as a measure to improve this important habitat for pollinating insects and small mammals. Arable farmers are rewarded for delaying the spraying and ploughing of winter stubble, a vital food source and cover for threatened bird species such as skylark and yellowhammer. Most of the farms in the scheme are now rodenticide-free and the use of non-toxic alternatives is encouraged in order to reduce and prevent secondary poisoning of predators such as barn owls, hen harriers, kestrels and stoats. Donal Sheehan emphasises that food companies must also accept responsibility for delivering more for nature by changing the way they pay for farm produce so that those farmers delivering more environmental benefits will be paid more. He says: 'If we can succeed in widespread implementation of these measures, I am confident that biodiversity levels can be improved significantly on intensive farmland.'

Hidden away on the shores of the picturesque inlet of Oysterhaven on the south coast of Cork, there is a very special area of coastal land where management is aimed wholly at improving biodiversity. Before these 20 hectares were bought by Emma Hart, about half were leased by a conventional tillage contractor, while the remainder – an area of mixed broadleaf woodland surrounding a lovely coastal lagoon – was too steep and wet to farm and thus largely unmanaged. Emma grew up in a cottage next to the land and spent her childhood exploring every corner, building dens in the woods, and she grew to love it here. Since she purchased the land in 2021, the management at Oysterhaven has centred on maximising biodiversity through a wide range of actions, drawing from the ever-expanding toolboxes of nature restoration, regenerative agriculture, agri-ecology, traditional conservation and rewilding.

On the land, all chemical inputs to the arable area were ceased and the majority was sowed with a species-rich mix of grasses, herbs and legumes to kick-start the soil biology. Some areas were sowed with a mix of species suitable for pollinating insects and winter feeding by small birds. These are grazed extensively in winter with rare native breeds of ponies and cattle while mob grazing is used in summer to move the animals around frequently, allowing the sward to recover quickly. The animals are kept at a very low stocking density which ensures a good balance between grazers and grassland, preventing a build-up of dead vegetation but also allowing the grasslands to flower and self-seed each year. The wild bird cover is grazed once a year in late spring (once the birds have benefitted from the leftover seed). It is then left to grow over the summer, set seed in autumn and provide food and shelter for birds and other wildlife throughout the winter. Emma says, 'We have seen a wonderful response from bird life in recent years, with a mixed flock of over 300 goldfinches, linnets, chaffinches and yellowhammers feeding both on the wild bird cover and the grassland.' In other areas, natural regeneration of trees is encouraged, with small-scale planting of native broadleaf trees in others. Emma says that in the few years since nature-friendly management began, it has been heartening to watch nature return, including sightings of over 70 species of birds.

Dramatic declines in farmland bird populations are a common problem across the whole of Europe and have been linked with increasingly intensive farming methods, removal of natural habitats and use of agricultural chemicals. Various attempts have been made to reverse these declines, but few have been successful. The RSPB in Northern Ireland completed a study showing that three key farmland bird species actually increased in number over a five-year period in response to the Environmental Farming Scheme (EFS) that compensates landowners for undertaking work to enhance biodiversity and water quality. The study assessed whether changes in the abundance of priority farmland bird species differed between 33 farms under the scheme management and 22 farms not subject to the management. It was conducted in County Down, one of the last remaining areas of lowland mixed arable farmland in Northern Ireland. On the farms taking part in the scheme, yellowhammer numbers went up by an impressive 78 per cent between 2006 and 2011. House sparrows were up 46 per cent and tree sparrows more than doubled in number in the five-year period.[48] The yellowhammer is a red-listed species (a bird of high conservation concern) which had been in sharp decline right across Ireland and this decline has continued in the wider countryside where measures are not in place.

Key options for farmers in the EFS were the provision of winter food crops for wild birds, retention of winter stubble, creation of arable margins and creation of pollinator margins. The RSPB farmland bird study, the first of its kind to be carried out on the island of Ireland, included face-to-face advisory work and showed that land management can improve the population status of farmland bird species. Kendrew Colhoun, then RSPB Senior Conservation Scientist, said:

> Our study was designed to evaluate whether the scheme options led to increases in the priority species and our conclusion was a resounding 'yes'. We see the EFS as a critical component as part of our work to maintain biodiversity across the countryside in Northern Ireland. Our study provides unequivocal evidence that these schemes can deliver for key species if the correct mix of EFS options (such as ones to provide summer and winter food and nesting habitat) are targeted to the right places and coupled with advice.

The recent decline in curlew in the border counties is understood to be largely due to loss of habitat through agricultural intensification and, in recent years, deterioration of habitats due to land abandonment and high rates of nest predation. Between 2011 and 2013, surveys over more than 100 square kilometres of suitable habitat showed just 18 pairs of breeding curlews at 14 sites in Donegal, Leitrim, Monaghan and Mayo. Farmers with breeding curlews on their land were invited to join the Breeding Curlew Grant Scheme. The farmers participating in the grant scheme all received training in the protection and enhancement of curlew breeding habitat. Additional habitat works have been carried out at some sites, including tree and scrub clearance to reduce cover for mammalian predators and perches for crows which might predate nests or young chicks. Other works include ditch reprofiling and establishment of a sluice, fencing works, rush clearance and liming. Due to the lack of curlew sites, the scope of the project was extended in 2012 to include sites important for other breeding waders – lapwing, redshank and snipe. Predator-proof fencing was erected at three of the last remaining breeding wader sites in Donegal.

Success was limited so a special Curlew Conservation Programme was established in 2017 to focus specifically on this threatened species in Ireland. It is funded and co-ordinated by the NPWS and includes surveys, nest protection, public and community engagement, combined with habitat restoration, maintenance, enhancement, and creation. Central to the approach are the owners of the land where Ireland's last remaining curlew breed. Nine geographical areas that are important for breeding curlew in Ireland were assigned locally based teams to work with local landowners and communities, adapting techniques and efforts to suit local needs. By 2023, 48 curlew pairs were recorded in these areas, down from 64 pairs three years earlier.[49] Despite huge effort and cost, the curlew is clearly in trouble and slipping towards extinction in Ireland.

By contrast, a focussed restoration project on the east coast of County Wicklow has recorded exciting results in restoring farmland habitats for breeding waders. At Cooldross, near Kilcoole, an area of low-lying coastal grassland is separated from the sea by a railway line that runs along the top of a shingle ridge. The land was used for silage and extensive cattle grazing until it was bought by the NPWS in 2006. The site was already important for wild swans and geese in the winter months but when waterbirds attempted to nest here in spring, the habitat conditions were not ideal and breeding success was low. Jason Monaghan, a Conservation Ranger with the NPWS, was assigned the task of restoring the site. He first commissioned a detailed ecological assessment and a management plan for the 40-hectare site, which includes a large lagoon.

Implementing the plan involved the management of water levels throughout the season by altering a sluice on the outlet channel to produce ideal conditions at the start of the nesting season. Next was the creation of 30 wet pools (or scrapes), runnels and other wet features. Reprofiling existing drains and shallowing the banks of the lagoon created a large amount of accessible muddy edge where waders and their chicks could feed. A 2.4-kilometre anti-predator fence was erected around most of the site to deter ground predators. Despite the electric fence, there has been a problem with fox and mink getting into the nesting area which reduced breeding success, requiring some targeted predator control. Cattle grazing and mowing regimes are delayed and balanced to avoid chick trampling and to maintain a short grass sward for chicks to feed and also to ensure suitable habitat for wintering geese and breeding lapwing for the following year. In places, this means strip mowing and grazing of certain fields. The use of agrochemicals on the land has been practically eliminated which is beneficial to the ecology of soil and water.

The results of the restoration work have been spectacular. From a low of five nests in 2017 the population had increased to 76 nests used by up to ten waterbird species of conservation concern in 2023. This included 50 lapwing nests which is probably the highest concentration of this red-listed species in the east of the country. The breeding success of lapwing has reached a point where the number of chicks produced is adequate to maintain the local population. Nests of redshank, ringed plover and shoveler show that the habitat is improving for these sensitive species too. Unlike many bird projects, this work is benefitting multiple species of birds, plants and insects by diversifying the habitat and using a management programme that is sensitive to the needs of all.

A trip to Amsterdam for an international conference in spring gave me a chance to see Dutch meadow birds in their classic habitats. I met up with the director of Vogelbescherming Nederland, the national bird protection organisation for the Netherlands, and we headed off across the flat lands of Holland, over canals and around marshes and polders. Intensive agriculture was everywhere, with large flat fields that had been reclaimed from the sea centuries earlier. I saw vast fields of crops without the familiar hedges that that we know in Ireland. The Netherlands is one of the most densely populated parts of Europe and yet they seem to have found space for some wildlife. Finally, we reached a wet polder still containing pools of surface water and here there were lots of waders – lapwings, redshanks and black-tailed godwits. In their rufous breeding plumage, the godwits looked strikingly handsome as the males rose up from the grass to display to each other.

Dr Jos Hooijmeijer, a researcher from the University of Groningen, has been studying these birds for some years. He says, 'We monitor around a thousand nests per year, but survival is low. We know from large-scale camera research that foxes, badgers and marsh harriers are mainly responsible for predation of eggs and chicks.' In North Holland, an area where the godwit is doing well on average, nature management appears to work well. In the Ronde Hoep, the Agricultural Collective has worked with farmers who receive compensation for nature-friendly measures. These include growing herb-rich grassland which is only mowed in July. Then the meadow bird chicks have plenty of time to grow up safely. The flowery land attracts so many insects that the chicks can find enough food later in the season. Lowering the water level makes the land more suitable for godwits returning from the south in the spring. This causes soil life to move upwards, making it more accessible to meadow birds. In the increasingly dry summers, the ground becomes very hard, which makes it difficult for the waders to feed by probing the soil. New dams on some farms actually provide a slightly higher water level, keeping the grassland moister. Droughts and predation of nests are a 'double whammy' for these sensitive birds.

Some birds represent flagship species for farmland conservation. Barn owls are fascinating birds that may breed surprisingly close to people and can suffer seriously as a result. Road collision casualties, poisoning by rodenticides and loss of nest sites in old buildings are just some of the pressures on these birds. Michael O'Clery remembers his excitement on his first visit to an owl nest site at Milltown, County Kerry. 'I watched a pair emerging from a hole, high on the gable end wall of the old building, and I was hooked. I started to look for other old buildings and ruins to see if there might be more in my area and quickly found a whole series of sites.' Michael then joined a Birdwatch Ireland survey team which showed that the main centre of the population of barn owls was indeed in the south-west of Ireland. The overlap with the distribution of the bank vole, one of the main prey species, was quite noticeable.

Other raptors, particularly kestrel and long-eared owl, began to feature in the survey team's work. For barn owls and kestrels, the team initiated nest-box schemes and published a booklet which they gave free to landowners, farmers and anyone interested. In recent years the response is such that it has been difficult for the team to keep track of just how many well-positioned, properly constructed boxes there are in the country. By 2023 they had recorded barn owls breeding in and using over 250 nest boxes, with this number increasing annually. In many cases, local people and small groups are now organising themselves to build and install nest boxes, with guidance whenever needed. I have recently erected a barn owl box in a tree on our own farm in County Wicklow. Barn owls are already breeding in this county after a long absence, and I am confident they will come here eventually. The box overlooks a traditional hay meadow that is full of small mammals including the recently arrived bank vole. These are all potential prey for the owls.

One of the rarest mammal species in Europe, with a stronghold in the West of Ireland, is the lesser horseshoe bat. I was once taken down into a cave near the famous Coole Park in south Galway by Dr Ferdia Marnell of the NPWS to see a winter roost of these curious animals. We crawled into a chamber off the main cave to find that the floor was covered in bat droppings. These had come from a small group of bats, tightly bunched together for heat, each one with its wings wrapped around it like a cloak. In Ireland these bats are confined to six western counties: Mayo, Galway, Clare, Limerick, Cork and Kerry. The reason for this may be that the west of Ireland is influenced by the Gulf Stream, so frosts are rarer and shorter in duration than in other parts of Ireland, resulting in greater availability of insects on which the bats can forage. The numerous hedgerows are ideal corridors for the bats as they move across open farmland. Also, caves are widespread in these six counties, particularly in Clare and Galway, and these provide suitable hibernation sites for the bats.

Research and survey work by the Vincent Wildlife Trust has shown that these bats prefer old stone buildings in the countryside for summer roosts, usually with natural slate roofs, because these offer a warm area, usually inside the attic, in which to rear their young. However, the bats are also found in sites that appear less suitable. Kate McAney has worked for the Vincent Wildlife Trust in Ireland since 1991. Under Kate's guidance, the Trust began acquiring buildings in Ireland used by the bats and currently manages 12 sites located in the western counties. All are summer breeding sites although, during mild winters, up to 1,000 bats hibernate in the Kerry sites, roosting in the ground floor areas. During summer 2023, a total of almost 3,500 lesser horseshoe bats were counted at Trust properties in Ireland, equivalent to about a quarter of the national population.

The course of action undertaken once a roost is leased or purchased by the Trust depends on the condition of the building. These are often in deserted farm cottages or out-buildings. In some cases, only minor works are necessary to ensure the building continues as a suitable breeding site but where the building is in an advanced state of disre-pair, major rebuilding and restoration work is undertaken. Abandoned farmhouses are widespread in the West of

Ireland where the rural population has declined dramatically in the last century and a half. Although the types of buildings acquired range from single-storey stone cottages to large two-storey stables, the aim of the renovation work is the same – to provide a structurally sound building that also meets the roosting needs of the bats.

One of the Trust reserves is at Rylane Cottage in County Clare. The bat colony using this stone building was discovered during an intensive 'door-to-door' survey of uninhabited rural buildings in County Clare in the late 1990s. At that time the colony consisted of approximately 50 bats that used gaps in a ground floor window and a chimney to enter and leave the cottage. Although the building was in a relatively good state of repair, the natural slate roof needed to be replaced and new ceilings erected to increase the temperature within the building. Using camcorders with infrared spotlights a total of almost 300 bats was recorded emerging from the cottage in June 2015. Overall, between 1993 and 2014, the Irish population of this species increased by between 60 and 97 per cent, although this should be viewed as a recovery rather than an expansion.[50] This clearly shows the importance of restoration work and the value of subsequent monitoring.

Farmland is the single largest component of our present landscape and much of the wildlife that lives here must adapt to living alongside farming activities that range from intensive cereal production in the east to marginal livestock farming on wet soils in the west.

The thousands of small mixed farms that supported large populations of Irish people in the past are long since gone, and with them some of the more sensitive wildlife habitats and species have steadily declined. Despite its green image, all is not well in the farmed landscape. The Countryside Bird Survey shows that there have been long-term declines in once-common grassland-nesting species such as skylark and meadow pipit. Even such widespread breeding birds as robin, song thrush and mistle thrush have been steadily decreasing over the last two decades with the removal and heavy trimming of hedgerows.[51] The Irish hare was once so common that it was seen every day in the countryside. But agricultural intensification, increases in larger, monocultural, mechanised farms and a move from hay-making to multiple silage cuts each year drove the numbers to an all-time low by the mid-1990s.[52] Insects in the countryside are among the best indicators of change. The Irish Biodiversity Data Centre reports an overall loss of 12 per cent in butterfly populations across Ireland in just the past ten years. The alarming declines in common bees and other pollinator species in farmland landscapes are thought to be linked to a decrease in suitable foraging areas and the widespread use of chemicals in agriculture.

The most promising nature restoration schemes in farmland today are those that are led by the farmers themselves. The landscape-scale schemes such as those in the Burren in County Clare and the Bride Valley of

County Cork all share the approach of subsidy payments based on results where the farmer has a menu of different measures tailored to individual farm habitats. The expensive agri-environment schemes of the past, such as the Rural Environment Protection Scheme (REPS) and the Green Low-Carbon Agri-Environment Scheme (GLAS), had very limited environmental benefits and certainly few of these have been sustained in the long term. These were largely designed by public servants who set out the range of options with little concern for the needs of individual farms or for how effective the measures were on the ground. The main target seemed to be to use European funds to subsidise non-profitable farming and keep rural people on the land. Of course, many individual landowners have taken actions on their own initiative to make their farms more wildlife-friendly. Several case studies have been described in this chapter. Each one of these depends on the energy and commitment of an enthusiastic farmer who understands the value of an environment that has a variety of habitats, each rich in wildlife.

Overall, there are many ways in which modern farming can be integrated with nature restoration, but each needs to be tailored to the special characteristics of the landscape where the farm is located. However, these projects on farmland are relatively few and scattered when compared to the areas that are managed intensively and based on EU subsidies. Where agri-environment schemes have been implemented, these are often based on short-term funding and continuity can be lost when this comes to an end. As the biodiversity crisis deepens, and we see the strong links to agricultural practices, there is a need to scale up successful approaches such as that of the Burren Programme to whole landscapes across the country. Liam Lysaght, Director of the National Biodiversity Data Centre, says, 'The irrefutable fact is that biodiversity loss cannot be adequately addressed without us finding ways that farmland can make space for nature.' Lysaght believes that the best way to do this is for all farms in receipt of Common Agricultural Policy (CAP) payments and other subsidies to be obliged to have a minimum of 20 per cent of their land area actively promoting biodiversity enhancement measures. He says, 'The farmers themselves should have the discretion as to how best to deliver this on their land, supported by empirical evidence that the measures are indeed delivering biodiversity benefits.'

Farming is deeply embedded in Irish culture and history. 'The crafts of arable farming, of animal husbandry and the home industries have done more to shape our instincts than the tramping of armies or the wrangling of kings,' wrote the folklorist Estyn Evans.[53] Even today, a large proportion of Irish people have farming in their family history. Centuries ago, Irish farmland was filled with wildflowers and the sounds of bees and birdsong. It was a richer and more vibrant place to live and work rather than a deserted industrial landscape. The restoration of nature on farmland will not only aid biodiversity and climate, but can also recreate this cultural value.

Rivers running free

Almost every day, I walk along the banks of the river that runs through our farm. From one day to the next, it can change from a bubbling stream to a raging torrent rushing off the hill to reach the sea. If I sit still for long enough, I might see a wild trout idling in one of the many pools. Occasionally a dipper flashes past, stopping to plunge into the water for the insect larvae that form its staple food. Footprints in the muddy edge show me that deer and badgers are coming to drink here at night and once I saw the distinctive outline of an otter emerging from the water at dusk. Unfortunately, the picture is not always so idyllic. On one occasion, I collected a bucketful of dead trout, killed by the release of slurry from a farm upstream. This confirmed for me that good water quality is the key to restoring our rivers.

The 40 shades of green for which Ireland was once famous depend on rainfall, usually greatest in the uplands where the peaty soils act like a giant sponge. As they release their load of water it trickles off in small streams and rivers, eventually arriving in lowland lakes and wetlands. Ireland's freshwaters and their biodiversity are currently in crisis. The number of rivers in Ireland that can be described as 'pristine' has declined to a handful. Many groups of freshwater invertebrates have some threatened species and up to 27 types of water beetle are threatened with extinction or are already regionally extinct.[54] In the 97 Irish rivers that support freshwater pearl mussels, all but one of these populations are threatened with extinction. These mussels, which can live for up to 120 years, are not reproducing and most of those remaining are ageing individuals. Invasive plant species can occupy the water column in lakes from bed to surface with serious consequences for native biodiversity. Some species, such as Nuttall's waterweed, curly-leaved waterweed and water fern, once established, can form exceptionally dense monocultures, excluding native species. Invasive invertebrate animals such as the zebra mussel, introduced game fish such as the dace and roach, and mammals such as American mink are also major threats to biodiversity in lakes. Overall, our waterways and wetlands are under serious pressure to survive.

Agriculture is the main source of pollutants including runoff of fertiliser and slurry, which is occurring with increasing frequency. Intensive forestry in a river catchment can also cause acidification and siltation in the water. Professor Mary Kelly Quinn of UCD has described the chemicals entering freshwaters as 'cocktails of pollution' and has shown that regional diversity is reduced in streams draining off intensive agricultural fields.[55] In the 1970s, several lakes in the north midlands were polluted by dumping of manure from industrial-scale pig farms. Alongside all this is physical damage to habitats through drainage and straightening of channels which causes massive losses of freshwater biodiversity. Many Irish rivers contain artificial barriers to fish movement, which often serve no useful function anymore.

The Intergovernmental Science-Policy Platform on Biodiversity and Ecosystem Services report of 2019 highlighted that freshwater ecosystems show among the highest rates of decline in condition of any of the habitats covered. Mary Kelly Quinn calls the loss of freshwater species 'an invisible tragedy hidden beneath the water surface'. Every species in such an ecosystem is linked to all the other species there. The key action needed in each case is to restore functioning ecosystems. This will not be easy and will not happen quickly as there are multiple stakeholders and getting them all to cooperate to pursue one objective is a difficult task. There are very few current examples in Ireland of where this has achieved positive results.

One way of controlling polluted runoff from farmyards is to create new settlement ponds that intercept the flow. These ponds are then planted with native wetland species such as reeds and bullrushes whose roots absorb the nutrients and prevent them from entering downstream watercourses. Feidhlim Harty is an expert in these techniques and he also supplies the plants needed by farmers to vegetate the ponds. He says,

Wetland-planted filter basins and ponds are more suited to impermeable soils, whereas tree-planted systems such as infiltration basins and wooded riparian buffer zones are more suitable for free-draining soils. In the ponds, lower-growing wetland plants form a biomass layer which provides physical, chemical and microbiological filtration.

In the Anne Valley near Tramore in south County Waterford, where dairy farming is the principal land use, a whole series of constructed wetlands have been created to deal with the washings of slurry tanks and silage pits. The project initially comprised reprofiling of the canalised stream and creation of a mosaic of medium to small water bodies. The ponds are generally shallow with abundant aquatic plants, although some have sections which are deeper and where vegetation is more sparse. In this area, three-quarters of all dirty water from farmyards generated within the watershed of the Anne River is channelled through these wetlands. Dr Rory Harrington has pioneered this concept on a river catchment scale with the active involvement of a group of landowners in the Anne Valley. He based the design for these constructed wetlands on hard science, but they were quite experimental at first.

A buffer of native trees or other semi-natural vegetation along the banks of a river also has the potential to absorb agricultural runoff and prevent this from entering the water. In the USA, this has been shown to reduce nitrate concentrations in shallow groundwater discharge from farmland by up to 90 per cent.[56] Riparian woods can also ameliorate the effects of many pesticides and provide dissolved and particulate organic food. This maintains high biological productivity and diversity in the rivers and streams that pass through them. Overhanging trees drop leaves and small woody material into the water, creating niches for invertebrates, and help to prevent oxygen depletion, while the shade cast by trees reduces water temperature, especially in hot summer weather. The roots of the trees and shrubs also help to bind the sediments in the riverbank, preventing erosion and silting of the riverbed. As part of the Native Woodland Scheme, the Irish government has introduced the Woodland for Water measure which provides grants and annual premiums to farmers and other landowners to afforest with native woodland along river corridors. Riparian woods were once common in river valleys, forming valuable natural corridors, but more intensive agriculture and associated drainage have removed many of these over the centuries.

Some farms are in the floodplains of large rivers where they suffer regular flooding and saturation of the soils. Michael Hickey runs a herd of Aberdeen Angus cattle on his 100-acre organic farm along the banks of the River Suir in County Tipperary in an area known as The Inches. The farm was semi-derelict when he took it over about 40 years ago. To prevent his herd from damaging the banks of the river he fenced off a riparian zone about 15 metres from the water's edge and allowed natural regeneration to turn this land into a woodland strip. The effect of the trees is to bind the soil and trap silt coming down the river, thus raising the level of the land and reducing flooding of the neighbouring fields. As a result, he gets a longer grazing season for his cattle and the farmyard manure that he spreads on the fields in the summer is absorbed better, preventing it from washing into the waterway. He finds that the band of mature trees also provides shade for his cattle in summer. He also comments on the abundance of birdlife on his farm and says that he often relaxes among the trees with nature all around him. He finds that this has enormous benefits for his own wellbeing. He has used these riparian zones as a nature-based solution to farming in wet areas.

A major concern at present is increasing water temperatures in rivers, due to the effects of climate change. Warmer water has implications for the survival of fish like salmon and trout as it leads to oxygen depletion. This issue was initially identified in Scotland, and Inland Fisheries Ireland are now assessing the potential impacts of the increasing temperatures on streams here.[57] The effects of this can be reduced by restoring natural shading along riverbanks with a strip of mature trees, so tree-planting has multiple benefits for freshwater and the biodiversity that it supports.

There are two types of drainage that are detrimental to rivers. One is land drainage which results in more rapid loss of water from the land to the river. The other is arterial drainage. The latter involves a permanent alteration of the channel to speed the flow of water to the sea, reducing flooding and lowering the general level of the river so that field drains can flow into it. Heavy machines are used to straighten, widen and deepen the channel as well as remove obstacles such as large rocks and bankside trees, protecting the banks from erosion with concrete or rock armour. The OPW is responsible for river drainage and it operates under an antiquated piece of legislation called the Arterial Drainage Act 1945.

The arterial drainage programme has led to some of the worst damage to river systems and their biodiversity throughout the country. Naturally meandering rivers with gravel bars and floodplains have been turned into straight, deep canals with neighbouring marshes and riverside meadows ruined. As a result, the habitats of sensitive species such as freshwater pearl mussel and white-clawed crayfish were destroyed, the rivers no longer had suitable spawning areas for trout and salmon and the flow rates were often too fast for slow-swimming fish. Today the OPW is responsible for 'maintaining' about 11,500 kilometres of channels in 33 catchments under this legislation. Local authorities are responsible for up to 5,000 kilometres of channels under various arterial drainage acts prior to 1945. About one-sixth of all lowland rivers in Ireland were dredged and drained by the state, so these are now extremely impoverished habitats for biodiversity.

This drainage work has had very significant long-term impacts on the capacity of Irish rivers to support salmon and trout. Even 60 years after the drainage was carried out, surveys have shown little physical recovery of the natural form in these channels and seriously depleted fish populations. Without restoration and enhancement, most of these rivers would not recover physically for hundreds of years or, in some cases, not at all. An Irish fisheries scientist, Dr Martin O'Grady, developed expertise in river restoration and oversaw the Irish government's investment in this work in the 1990s. He worked for the Central Fisheries Board for decades and then with Inland Fisheries Ireland until his untimely death in 2018. O'Grady devoted much of his professional life to researching how rivers and lakes could be rehabilitated to improve the quality of fish stocks. During numerous fisheries surveys, he undertook over three hundred flights with the Irish Air Corps. With this bird's eye view, he could clearly see the damage that had been done by straightening and deepening rivers, removing gravel beds and draining adjacent wetlands. Among his significant contributions was a manual entitled *Channels and Challenges* which sets out in great detail the measures needed to enhance rivers for the most sensitive fish species including salmon and trout.[58]

O'Grady used a number of techniques that had been employed in other countries to enhance salmon rivers. The first was the stabilisation of riverbanks using a combination of logs, rocks and conifer tree tops, which he called his 'Christmas tree technique'. Slowing down the flow was the objective, allowing suspended silt to settle out around the obstacles. This silt is rich in nutrients and quickly becomes vegetated. With stockproof fencing to prevent cattle causing further damage when they enter the water, the tree tops rot after a few years, leaving the logs beneath them buried in silt. In a natural river channel with an average width of five metres, a pool would develop every 25 to 30 metres down the river. These pools are vital for salmon and trout to rest during their marathon migrations upstream to spawn in the headwaters. Many such pools were destroyed during the arterial drainage programme. To slow the flow and allow pools to develop again, a series of shallow weirs were created. These used a variety of materials including timber and stone each with at least one notch in the barrier to allow water to continue to flow during low water levels.

Climate change is predicted to bring much lower flows in most rivers in the east of the country, although there are likely to be more frequent exceptional rainfall events that can cause major flood damage in towns and cities when the rivers overflow. In such areas, a restoration technique is the creation of two-stage channels involving a deeper part which would still contain water in low flow, while the shallower areas would be covered in flood conditions. This has occurred several times on the River Dodder at Ballsbridge near Dublin city centre. Here, the OPW installed major flood defences to protect local properties. Dr Ken Whelan, working with Alan O'Sullivan of Rivus, recently used one of his innovative designs to reinstate natural gravel and cobble beds into the river thus restoring the natural sinuosity of the river channel and slowing the flow.

Silting of the gravel beds in rivers is a major problem for spawning fish and other aquatic life. This often occurs as a result of excessive erosion of riverbanks due to unrestricted access by livestock to the water's edge. Their hooves break up the vegetation binding the soil and floods undermine the bank causing slumping into the water. This can be stopped by fencing to prevent livestock gaining access to the riverbank, but it requires some extra measures to restore the vegetation and stability to the river. To find out more, I met Gareth Pedley, a conservation officer with the Wild Trout Trust, a charity working across the UK and Ireland to help anyone interested in restoring rivers, lakes and their wildlife, including wild trout. He has provided expert advice to landowners along the Culdaff River, close to Malin Head, County Donegal, one of the most northerly rivers in Ireland. Here, there was block failure of the riverbank caused by a lack of bank stability, due directly to livestock grazing and denuded bankside vegetation. The work on the riverbank was undertaken by volunteers from the Inishowen Rivers Trust and the Culdaff Community Angling Club.

Cut branches were used as bank protection to stabilise eroding areas, particularly protecting the toe of the bank. This allowed time for trees and vegetation to become established and helped to consolidate the bank and reduce erosion to a more natural rate. As livestock grazing and physical damage to the banks, including trampling of the bank top, were a significant contributing factor, buffer fencing was also a key aspect of the solution. Willow saplings were planted to create fast-growing live protection along the riverbank, perfect for locations where access is not required. The end result resembled a hedge of live willow, significantly slowing the flow of water through it. In addition, dead wood material was added to create roughness along the bank, providing protection from the erosive forces of the river and encouraging deposition for as long as the material persists. This created a more stable base into which various other species (planted and naturally propagated) can become established. However, Gareth says,

> In some artificially realigned channels, habitat quality may actually benefit from the channel being allowed to erode laterally to restore former meanders so bank protection in those situations may not be beneficial. The prescription really is on a case-by-case basis, so the actual work that is entailed and how it's delivered can also vary.

Glenasmole in the Dublin Mountains holds the Bohernabreena Reservoir, providing an important supply of domestic water to the growing city of Dublin. In 2016, almost 2,000 hectares in the upper valley were acquired by the state to be added to Wicklow Mountains National Park. The area has suffered from peat erosion and overgrazing by sheep and deer. It also holds numerous tributaries of the River Dodder, but some native trees have survived in the gullies here. With the help of many volunteers, the National Park staff have started to plant additional native trees in these areas including rowan, hawthorn, birch, hazel, oak, blackthorn and holly. There is some evidence of naturally regenerating trees as well, so small fencing exclosures have been erected near the streams to protect these and the newly planted trees from grazing animals. Bands of trees beside the water help to filter out peat from further upstream thus improving water quality in the river and reservoir downstream. Vast amounts of peat are being lost from the Dublin Mountains with huge areas of bare peat between Seefingan and Kippure upstream. Initial work to stem the erosion will be carried out on Red Hill with stone and timber dams, heather brashing and reseeding with fencing to exclude grazing animals.

The River Vartry flows through the centre of my local village in County Wicklow. For centuries there was a steep stone weir just upstream of the village, built to provide water power for a corn mill at the famous Mount Usher House. I have watched a kingfisher perching on an overhanging branch here waiting to plunge into the pool above the weir for a small fish. But, with the mill abandoned in the late 19th century, the weir no longer had any economic value and it formed a significant barrier for fish moving upstream, especially during low water periods. Then, following one major flood, the structure broke up and floodwater poured through the weir as if released from captivity. In effect, the river was returning to its natural bed.

Irish rivers are heavily fragmented by weirs, dams, sluices, culverts, bridges and other artificial barriers, breaking up their smooth flow from source to the sea. Fragmentation is one of the greatest threats to freshwater ecosystems because it interrupts fish migration, blocks the movement of other aquatic animals within the channel, alters the flow of sediment, affects habitat diversity and impacts overall biodiversity. These structures cause the loss of many natural features and add to the other pressures on biodiversity such as water pollution, channel alteration and water abstraction. Inland Fisheries Ireland (IFI), the government agency that aims to protect our fishery waters, has been monitoring the barriers on Irish rivers. They have also been joined by colleagues in the Local Authority Waters Programme (LAWPRO), a number of whom have been trained to help record barriers. As a result, they have identified over 73,000 potential barriers on the Irish river network. To date, they have surveyed almost 22,600 of these structures, recording their construction and dimensions and their potential to prevent fish movement and migration. Removing some of the obsolete ones is the next step but this has proved to be difficult to implement in Ireland.

In many other countries, removal of barriers is already allowing fish to return to spawning grounds in the headwaters from which they were excluded for centuries. In Europe, almost 5,000 dam removals had been recorded by 2020, from 11 countries. The EU 2030 Biodiversity Strategy calls for greater efforts to restore freshwater ecosystems and the natural functions of rivers. It sets a target to make at least 25,000 kilometres of rivers across Europe free-flowing again by 2030, primarily by removing obsolete barriers and restoring floodplains and wetlands. In Ireland, the migration of salmon, sea trout and lamprey from the ocean to their spawning grounds in the headwaters of many rivers has been interrupted by barriers such as weirs, culverts and hydroelectric stations. Seriously threatened eels are also badly affected by these barriers. 'Fish passes' and 'fish ladders' with shallow pools can make a difference, but these are expensive to construct and only allow a fraction of the fish to pass. The removal of obsolete barriers would open up a huge number of tributaries to migratory fish and many other species and would cause little impact on the national electricity grid. It may also help to reduce flooding of properties upstream of the barrier.

There are other ways to overcome river barriers where a weir is still necessary. In East County Galway, the River Bunowen, a tributary of the River Suck, was diverted centuries ago to power a now derelict mill at Ahascragh. Weir gates were constructed on the natural river channel and these were closed in low water conditions to maintain the flow in the headrace above the mill. In normal conditions, water powered through the weir gates creating a waterfall that was impassible for fish. Here, IFI designed an innovative fish ladder that involved a series of shallow steps across the full width of the river below the weir to bring the water level gradually up to the top of the barrier.

This was constructed in 2018 with 14 rows of rocks across the river at intervals extending for 90 metres downstream. The change in water levels between the steps was only about 15 centimetres and the water flow was sufficiently slowed to allow migratory fish like salmon and trout to pass upstream.

The East Wicklow Rivers Trust also has plans to remove or modify several barriers including one at Ballinglen Bridge on the Aughrim River. This voluntary body covers an area of over 1,300 square kilometres stretching from Bray to Arklow, including the Vartry, Avoca and Dargle catchments and all the rivers, streams and lakes that flow into these. The establishment of rivers trusts throughout Ireland is an encouraging sign that things may be changing. Regional Trusts have been established in at least 18 parts of the country from Inishowen in north Donegal to the Bandon River in Cork. These are independent community-led charitable organisations, focussing on education, water management advice and practical conservation work, from source to sea – improving land, rivers and wetlands at a catchment or river basin scale. They hold regular riverside workshops, showing anglers, landowners and ordinary citizens how to monitor the water quality in 'their' rivers, using simple methods such as sampling of invertebrates.

Rivers and their associated wetland habitats are among the most seriously threatened habitats in Ireland due to a combination of water pollution, arterial drainage, artificial barriers and bankside erosion. There are plenty of examples of river restoration methods that have been successfully used in other countries. One of these is restoring the floodplains on lowland rivers by creating meanders where none exist. This technique, often called 'rewiggling', has been used in a number of European countries. At Goldrill Beck in the English Lake District, nearly two kilometres of new channel were dug to reconnect this river with its historic floodplain, moving it away from a straightened channel, thus reducing flooding downstream and restoring habitats for fish and invertebrates.

Unfortunately, the authorities in Ireland have been slow to implement these restoration methods on our rivers. This is partly because of the multiplicity of organisations, state and voluntary, that are involved and a lack of central coordination. One question I am frequently asked is who a person should notify if they become aware of a pollution problem, and the answer is rarely clear. The causes of declining river quality are many and the public are becoming more aware of these issues, but the losses of freshwater biodiversity are less well known. With their feet in the water on a regular basis, anglers are usually highly sensitive to any changes of this sort and they are often involved in practical efforts to restore their favourite stretches of river. But the crisis is deepening year on year and a major effort will be required by all sectors of the community to return our rivers and their wildlife to their once-healthy state. There is a legend that the goddess Sionnan, after whom the River Shannon is named, ate the Salmon of Knowledge and thus became the wisest person in the world. We need some of this wisdom now, to help restore our rivers and the benefits that they bring to everyone.

Peatland rehab

I have to admit that walking across a bog is a slightly alien experience for me, as there are no trees for shelter and the ground is wet and occasionally treacherous. More than once I have slipped and fallen into a drain or bog pool. For many Irish people, bogs represent a wasteland, useless for farming and only good to be dug up and burned to keep them warm in winter. As the population began to grow in the 16th and 17th centuries, coming at the same time under the control of foreign powers, 'reclamation' of the bogs for agriculture became the focus. First to be cut away were the bogs near Dublin to supply fuel to a burgeoning city population. By the 18th century, there was a growing interest in bogland reclamation among the landed gentry, who wanted to make their estates as productive as possible in order to generate more rental income from their tenants. In 1731, an Act of Parliament was passed to encourage the reclamation of bog and other 'barren and waste land'.[59] But intact bogs are far from barren and hold a rich diversity of life.

The plants in a wet bog are its most important feature, although a more difficult environment is hard to imagine. The soil in which they grow is highly acidic and composed of partly decomposed remains of earlier plants. The reason they do not fully decompose is that the peat lacks oxygen due to waterlogging and the bacteria and fungi that normally break down cellulose cannot survive here. This means that the nutrients that living plants require are locked away in the peat. Many of the bog plants deal with this by forming associations with fungi called *mycorrhizae* in their roots. Perhaps the best-known plants of bogs are the heathers whose flowers give a characteristic purple hue to the habitat in the autumn. They occur mainly on drier, better-drained hummocks above the wet surface. In wetter bog, the mosses are often dominant with the greens and reds of the *Sphagnum* group constituting the most important. They hold large quantities of water, like a sponge, and their growth and layering is what causes bogs to grow. In these conditions, colourful and specialist plants grow – sundew, asphodel, bogbean and milkwort with sheets of white bog cotton flowers waving in the breeze. These plant communities have survived here for thousands of years.

As Ireland emerged from the grip of the last ice age, the flat midland area would have looked a bit like Greenland today – a network of braided rivers and partly frozen lakes filled the landscape as far as the eye could see. By the time the first humans appeared over the horizon in the Mesolithic period, vegetation was already well established and the shallow lakes would have been surrounded by dense bands of reeds and wet woodland. As the millennia ticked by, the dead vegetation built up in the beds of these lakes and, with a damper climate developing in Ireland, conditions were right for the growth of raised bogs in these lakelands. For ancient peoples, the peatlands were cherished for their preservative powers; many large lumps of butter, buried to preserve them, have been found in the bogs of Ireland. The peat also preserves human remains and these 'bog bodies' have provided information of great interest to archaeologists on how people lived here in the past.

Arthur Young was an English writer who travelled through Ireland, later publishing his *Tour in Ireland, with General Observations on the Present State of that Kingdom in 1776–78.* He described unbroken tracts of bogland: 'containing thirty-five square miles or twenty-two thousand four hundred acres. What an immense field of improvement! Nothing would be easier than to drain it (vast tracts of land have such a fall), that not a drop of water could remain.' Turf-cutting was, for centuries, the main source of fuel for many communities in the midlands and west of the country and this resulted in damage to raised bogs in particular as the water began to flow out from the margins and the domes in the centre sank and contracted.

The peatlands once covered up to a sixth of Ireland, a higher proportion than any other European country, with the exception of Finland. They can be divided into raised bogs (mainly across the midlands), blanket bogs (both in mountains and at low levels in the West of Ireland) and fens (which are widespread but generally small). When the time came to survey and protect some of the remaining raised bogs for conservation, there was so little left intact that the results were a shock. What remained was damaged, punctured, drained and drying out. No bog in the country was unaffected by the turf spade, the turf-cutting machine or 'hopper' and the industrial harvesters of Bord na Móna, the state peat production company. It is estimated that there had been a 99 per cent loss of the original area of actively growing raised bogs with only about 1,650 hectares of the remaining 'intact' high bog now classified as living 'active raised bog'.

Bord na Móna began mechanised peat harvesting in the middle of the 20th century and at its peak of production supplied over one million tonnes of peat every year, much of it to peat-fired power stations. By the late 20th century, little remained of the raised bogs as the peat had been mined and burned or sold off to the horticultural industry as 'moss peat'. Like many other people, I was unaware of the almost complete loss of the original peatlands in the centre of Ireland. Surprisingly, my awakening was due to the campaign of a group of ecologists from the Netherlands called the Dutch Foundation for the Conservation of Irish Bogs. In the 1980s, these people collaborated with the newly formed Irish Peatlands Conservation Council (IPCC) to bring a number of Irish politicians, public servants and ecologists to see the last bog remnants in the Netherlands, and I was lucky to be on the trip. Here the Irish group walked over the surface of a much-prized bog to view the emergency action that had to be taken to keep water flowing through the surface layers, much like a hospital patient on life support. All around this area, intensive agriculture, small towns and motorways stretched away into the horizon.

Back in Ireland, the campaign gathered pace with attention focussed on the last intact raised bogs in the midland counties. Under considerable political pressure from the European Union, the best remaining sites were eventually designated as Special Areas of Conservation (SACs). This produced a negative reaction from many local communities who regarded turf-cutting as a traditional right and resented the interference of government and the EU. From then on, serious differences arose between conservationists and local communities including the turf-cutting contractors who made a living from extracting – essentially mining – the peat. The opposition to the turf-cutting restrictions generated significant political pressure and the government initially backed off and deferred any enforcement action for a decade. A bog purchase scheme was put in place from then and large areas of bog and many smaller plots with turbary rights were bought by the state. Finally, when forced by the European Court of Justice, the government instigated a massive compensation scheme making annual payments or supplying eligible people with alternative sources of winter fuel. The cost of this scheme consumed a large part of the total budget of the NPWS for a number of years and it continues to this day. Today, however, there are 53 protected peatland SACs, 75 Natural Heritage Areas (NHAs) and about 30 other non-designated smaller bogs. Very few of these have had any industrial harvesting.

In the 1980s, when I first visited the midland bogs, most of them were vast wastelands, stripped of their peat and ploughed in long straight lines by huge milling machines. In the area around Lough Boora in County Offaly, Bord na Móna tried planting the remnant peat and mineral soils beneath with conifer plantations, but this was not commercially viable. When the pumps that had drained most of the bogs were removed, the depressions in the landscape naturally flooded again, forming (or reforming) huge areas of shallow lakes which were rapidly colonised around their margins by reedbeds and birch-dominated woodlands. In 1996, a group of peat workers had another idea for the use of this recovering landscape. They saw the tourism potential of these new habitats and established the Boora Parklands where the more natural landscape could be enjoyed by increasing numbers of visitors. Cycling and walking routes were laid out to allow visitors to absorb the unique atmosphere of this special place. Birdwatching hides were established to help visitors get close views of flocks of ducks, geese and swans that had returned to the expanses of open water on the flooded peatlands. The parklands now attract over 100,000 visitors a year, just one indication of the value of this outdoor amenity. One summer, I hired a bike to cycle around the level paths and explore a large area while getting some much-needed exercise. The large wetlands fringed by reedbeds and woodland offered stunning views and reflections from the water surface.

Since the 1990s, Bord na Móna has been actively carrying out research into restoration and rehabilitation of cutaway bogs for the purposes of promoting biodiversity. Restoration involves those peatlands that have been drained for cutting but never harvested. Drains are blocked to rewet the bogs and return them to peat-forming conditions. A key to the growth of bogs is to reestablish a healthy coverage of *Sphagnum* mosses. This also stops further release of carbon dioxide and, in time, enables the bog to absorb this 'greenhouse gas' from the atmosphere. So far, the company has restored 3,500 hectares of bog, changing the trajectory of these damaged bogs back to peat-forming conditions, with only about 500 hectares more left in good enough condition to restore. Where most of the peat has previously been cut away, these former peat production areas cannot be restored to their original condition and must be rehabilitated instead. With a combination of natural regeneration and targeted rewetting, they can return relatively quickly to species-diverse areas. These formerly bare cutaway bogs are being transformed into wet woodlands, grasslands, reedbeds, fens and open-water habitats. Under its Peatlands Climate Action Scheme the company aims to carry out enhanced rehabilitation measures on 33,000 hectares or more than one-third of its landbank.

On a sparkling day in autumn, Bord na Móna's chief ecologist Dr Mark McCorry took me to visit several of the bogs currently in transition. Derrinboy Bog in west Offaly was first surveyed in the 1980s when this site was classified as intact bog. Subsequently, it was drained and some of the top layers were harvested for horticultural peat moss but at least two to four metres of peat remained unexploited. Looking out across the flat landscape, it was clear that some of this is acid in nature, known as 'red peat', while the remainder is more alkaline, being formed as fen, and is termed 'black peat'. Industrial peat extraction ceased here in 2019 and the bog was rewetted in 2022–23. The rehabilitation work here involves creating a network of raised banks enclosing cells that will flood naturally. The objective is to achieve water depths of not more than ten centimetres which are the conditions required for the growth of the all-important bog-building moss *Sphagnum*. Workers, previously employed in the industrial harvesting of peat, have been re-trained to undertake this delicate job of rebuilding the bogs and there is now a skilled team using over a hundred excavators and bulldozers for this work. It is obvious that the restored or rewetted bog will look quite different to the original intact peatland, but the revegetated surface will at least reduce the loss of carbon dioxide to the atmosphere and, in time, provide a habitat for a return of typical bog plants and animals.

The second site that I visited with Mark McCorry was Derrybrat Bog near Ferbane, which already looked a lot more like natural habitat. Most of the peat had been removed by the time industrial production ceased in 2018 but drain-blocking and creation of raised banks here between 2016 and 2023 was already rewetting the site. The result is extensive open water and a mosaic of vegetated areas with reedbeds, fen, wet woodland and other habitats developing. The underlying marl or lime exposed by peat extraction makes the water here more alkaline than acid. This produces water chemistry that suits the growth of stoneworts, unique submerged plants that form a calcareous crust. Other unusual plants that have returned to these rehabilitated bogs include the round-leaved wintergreen, a new species for west County Offaly and now found in up to six sites in the county. The rare marsh fritillary butterfly has also returned and has now been found in around 30 of these rehabilitated peatlands. Once the open water becomes a permanent feature, it attracts flocks of wintering wildfowl such as wigeon, teal and whooper swan, while threatened waders such as snipe and lapwing have started to breed again in these new habitats.

One of the exciting bird species that has benefitted from the rehabilitation of midland bogs are common cranes, large, graceful birds that were once common in Ireland but disappeared many centuries ago. They were a popular food item for people in earlier times and the bones of cranes have been found in archaeological excavations at Neolithic/Bronze Age sites in Dublin, Wexford, Roscrea and at Lough Gur in Limerick. The 12th century naturalist Giraldus Cambrensis wrote in his *History and Topography of Ireland* that 'cranes are so numerous that in one flock alone you will often see a hundred or about that number'.[60] Their vulnerability to predation by foxes and the historical draining of wetlands resulted in their final demise sometime between 1600 and 1700. The exact date of extinction in Ireland is unclear but may have been linked with over-hunting for food in Norman times.[61] The crane still breeds in other parts of Europe and a small population has returned to nest in wetlands in parts of Britain.

In May 2021, Bord na Móna announced that a pair of cranes was nesting on a rewetted bog in the midlands. There were two previous breeding attempts at the same location which were ultimately unsuccessful. While two chicks did hatch from this first Irish nest they disappeared shortly afterwards, probably due to predation. Then, in 2022, came the exciting news that the same pair had produced two more chicks which were now flying, making this the first successful breeding of this species in Ireland for about 300 years. By the following year, another chick had fledged and cranes were already prospecting other rewetted peatlands during the breeding season, so it looks like they are here to stay. Crane nests float amongst emergent wetland vegetation such as reeds so the new habitats, created by the ongoing work of Bord na Móna in rewetting some bogs and rehabilitating others, are ideal for restoration of these long-extinct birds.

Another constructive approach was taken in the Living Bog project, a raised bog restoration effort which ran from 2016 to 2021 at 12 bogs across seven counties, led by the NPWS with funding from the EU LIFE Nature & Biodiversity Fund. The project aims were to restore active raised bog in at least some of Ireland's SAC network in seven counties, improving over 2,600 hectares of important raised bog habitat, the equivalent of almost four times the size of Dublin's Phoenix Park and 18 per cent of the national high bog area. Over this five-year period, it was hoped to bring back a habitat which could support hundreds of plant species and many of Ireland's rarest animals including birds and insects. The scope of the project was huge. Over 200 kilometres of drains associated with turf-cutting were to be blocked on high and cutover bog areas using 15,000 peat dams and almost 1,000 plastic dams to raise water levels and rewet the bogs. Infilling of drains, removal of certain invasive plants, fencing and walkway improvements, fire plans and amenity provision formed just a small part of the overall restoration project. Restoration of active raised bogs is an ambitious, complex enterprise, recreating the hydrological and ecological conditions under which super-absorbent *Sphagnum* moss habitats will form new peat.

The Living Bog project also aimed to secure local community co-operation and foster a greater national understanding of the importance of Ireland's raised bogs and the significance of SACs through a variety of outreach and public awareness projects. Local involvement at two bogs, Killyconny Bog and Carrownagappul Bog, was particularly successful. With promotion of the sites during (and after) project work, it was hoped there would be a significant socio-economic spin-off for local communities as a result of this project. The amenity value of most of the sites is very favourable and, although the bogs will not be used for fuel, they can still play a key role at the heart of local communities. From its base in Mullingar, the five-person project team was led by Jack McGauley, Project Manager. I asked Jack about the main challenges for the team.

According to him the sheer scale of the works was quite daunting and the timescale (five years) was very short. The largest of the sites in this project was Clara Bog Nature Reserve, probably the finest remaining example of a raised bog in Ireland today. It covers approximately 840 hectares, of which 443 is uncut high bog with the remaining 393 hectares mostly cutover as a result of previous turf-cutting and peat removal. The objective of this project was to restore at least 180 hectares of active raised bog at Clara by 2020. This is roughly a tenth of all the active raised bog that remains in Ireland. To block drains, the project team removed surface vegetation or 'scraw' from either side of the drain where the dam was located. They then cut a 'key' in the drain and flattened or compacted the peat. They took peat from two 'borrow pits', located side-by-side. Small quantities of peat from the pits were placed into the drain, compacting it in the process. They built up the dam and placed the scraw back on top of it. The borrow pits were then reprofiled, breaking down the central vegetation mound and putting back other scraw. This gave the borrow pit a good profile and encouraged new vegetation growth. In previous years, borrow pits would have been left as a type of 'bog hole'. This new method encourages the quick recolonisation of vegetation in the borrow pit.

Now, several years after the project ended, Jack McGauley says that the slow rate of progress was also a challenge. It takes years to gauge if water levels are rising and to measure results in terms of ecology. On previous restoration projects, including works done on Killyconny Bog SAC, County Cavan, in the early 2000s, it took as long as eight years to get a true picture of the effects of the work, and up to 30 years may be needed after appropriate restoration works are undertaken on cutover areas for a bog-like habitat to develop. Encouraging results were recorded by the project ecologists, who found peat-forming habitats on the cutover. The techniques developed by this project are now routinely used by the NPWS. Since 2018, this state agency has restored approximately 2,500 hectares of habitat within the raised bog network of designated sites (SACs and NHAs). While the industrial exploitation of Ireland's bogs lasted less than a century, it will take a lot longer to restore some of the surviving examples to full health.

In 2004, the state forestry company Coillte began an EU LIFE Nature Programme to actively restore 571 hectares of raised bog on 14 midland sites in counties Galway, Roscommon, Longford, Westmeath, Meath, Cavan and Laois. All of these sites are designated as candidate Special Areas of Conservation (cSACs) under the EU Habitats Directive and provide habitat for a range of nationally important rare plant and animal species. It is estimated that Ireland contains 50 per cent of the intact oceanic raised bog systems remaining in Europe. In 2007, Coillte also completed a five-year EU LIFE Nature Blanket Bog restoration project. Over the course of the project, they used various restoration techniques such as tree removal, felling of trees to waste and blocking drains to rewet previously drained areas. This project resulted in almost 2,000 hectares of blanket bog being restored. As well as the operational work, the project included public awareness programmes consisting of information days, distribution of project brochures and the publication of a project results booklet. Another Coillte project, 'Demonstrating Best Practice in Raised Bog Restoration in Ireland' was jointly funded by the EU Commission, the Department of Culture, Heritage and the Gaeltacht and Coillte under the EU LIFE Nature Programme. The project focussed on the restoration of 636 hectares of raised bog habitat on 17 Coillte-owned sites within the Natura 2000 Network and in NHAs. This project implemented best practice restoration techniques developed in Coillte's previous Raised Bog Restoration Project.

Northern Ireland's second largest surviving area of intact raised bog is located south of Maghera, County Derry. It is not far from the birthplace of Seamus Heaney, one of Ireland's most celebrated poets. Several of Heaney's poems speak of the importance of bogland in the local traditions of this area. Ballynahone Bog is the largest of Ulster Wildlife's nature reserves and has an incredible range of bog flora, including nationally rare bog rosemary, *Sphagnum* mosses and liverworts. The bog and surrounding birch woodland are very rich in birds, dragonflies and butterflies. Priority bird species using the bog include skylark, merlin, breeding common snipe and a small heronry.

Some years ago, the site was acquired by Bulrush Horticulture with a view to harvesting the peat for supply to professional and amateur growers. This company put in a large central drain and about 50 side drains across the southern half of the bog. The local outcry about the potential destruction of this unique peatland led to a compulsory purchase of the bog by the Northern Ireland government. In the following years, the Northern Ireland Environment Agency (NIEA) blocked many of the drains, but these dams were not functioning properly. Ulster Wildlife then took a 99-year lease of the site in 2000. This conservation body commissioned a detailed Lidar survey to map the fine contours of the bog surface. This led to the installation of several hundred plastic dams, each carefully located to ensure that the drainage of the bog was arrested. This work was supported by the NIEA Challenge Fund. To monitor the effect of this work on water levels, a number of data loggers were installed in the bog. Hydrology is key to peatland management and restoration. Unless the natural water retention capacity of the peat can be restored it will not provide the right conditions for the growth of *Sphagnum* mosses which are the vital bog builders. The growth of peat may take centuries to achieve, but at least the damage has been stopped and conditions are now favourable for conservation of the sensitive species that use this important bog. Sadly, nitrogen deposition, which changes the vegetation, is a real problem at Ballynahone Bog, given its location in the heart of industrial farming in Northern Ireland. This cannot be addressed on the bog itself and will require significant change of land use in the surrounding area.

Abbeyleix Bog is situated on the southern periphery of Abbeyleix town in County Laois. It encompasses an area of almost 190 hectares of diverse habitats including recovering raised bog, cutaway bog once worked by local people, lagg around the edges, wet carr woodland, alkaline fen and wet meadows. This was part of the de Vesci family estate since the 1700s. In 1865, the bog was bisected by a railway which had dramatic effects on the hydrology of this wetland. It was compulsorily acquired by Bord na Móna in 1987 and, two years later, over 60 kilometres of drains were cut through the bog in preparation for harvesting the turf. But the local community here would not accept active damage to 'their' peatland. The Abbeyleix Bog Project was established in 2000 to conserve and protect the bog which was then threatened with harvesting for peat moss. There was a local protest to prevent machines from entering the site. Laois County Council then requested that a formal environmental assessment be carried out.

A turning point came in 2009 when bog restoration was carried out by Bord na Móna and NPWS and all the drains were blocked as part of an intensive drain-blocking programme, with rewetting the bog as the main objective. More recently, bunding has been used on the cutaway bog to trap water in cells. Following negotiations between the parties, a lease was signed in 2010 when Bord na Móna handed the bog over to the local community to manage for a period of 50 years with a primary focus on conservation. The project group developed a Conservation Management Plan in 2015, following on from a Business Management Plan in 2014, to maximise the ecological benefits to the local community in terms of education, employment, income and recreation. The Abbeyleix Bog Project is an open-access amenity developed by volunteers to benefit all.[62] As well as restoring the high bog, a local team from the community has also been clearing bracken to create more species-rich grassland. The invasive shrub rhododendron is spreading in some areas and this requires constant work to remove new plants that try to colonise the bog.

I walked around the bog in August when the heather was in full bloom and many butterflies were on the wing in sheltered sunny glades that have been cleared in the bog woodland. My guide was Chris Uys, a local Abbeyleix resident. He told me that one of the key objectives of the Abbeyleix project is to make the bog available for educational purposes in order to promote conservation of this unique habitat. Typical of peatlands, there are pools, open drains, rough ground and very wet and unstable conditions underfoot. A boardwalk has been built by volunteer effort and this facilitates access to the high bog

– a habitat that is otherwise difficult to enter. The old railway embankment has now become the main walking route, allowing visitors to reach right into the heart of the bog. This is an outstanding example of community-led action to protect part of the local heritage from destruction and make it available to a wider population.

A template for high-quality visitor facilities and interpretation of the peatland ecosystem was evident at another site I visited at Lullymore Bog in County Kildare. Part of this bog was handed over by Bord na Móna to the local community and has been developed as an attractive local tourist venue with a visitor centre and nature trails, providing an alternative livelihood for families that once depended heavily on the turf extraction industry. The opportunity to use these sites for education and interpretation has been taken with excellent involvement of local schools, a measure that should engage whole communities into the future.

Killaun Bog, close to Birr in County Offaly, has been fully embraced by a local school. The bog was severely damaged over centuries of exploitation, first by private owners and latterly by Bord na Móna. Most of the nine metres of peat depth was removed exposing the underlying mineral soils which were deposited by glaciers in the last ice age. Since industrial peat cutting ceased, it has begun to revegetate naturally. A local environmental scientist, Dr John Feehan, has eloquently described the natural history of the bog in considerable detail and has been a driving force, over the last 30 years, in securing its long-term protection. One of the unique aspects of this site is that over 40 hectares of the bog is owned by the local St Brendan's Community School. John and the teachers here have been active in introducing the students to the wonders of nature, thus sowing the seeds for the involvement of future generations of local people in the restoration of the bog and confirming its value to the community. In 2021, John published a book, *When the Nightjar Returns*, on the natural history and human story of Killaun Bog. This led me to meet the author to discuss his vision for the future. In the book, he outlines various scenarios for the restoration of biodiversity here. He says, 'Where residual peat remains on mineral soil the optimal strategy in the short term for biodiversity and carbon sequestration would be to remove most of the remaining peat so that trees and other plants can access a more adequate nutrient supply.' He envisages an alternative approach for areas where the natural processes of recolonisation have been active longest. 'These provide a glimpse of the possibilities for the spontaneous return of a high level of biodiversity in equilibrium with the changing conditions.' But bogs take time to recover. He says, 'The vision of the Greater Killaun Project will not become reality in one or two decades but over centuries, becoming ever richer as time goes by and nature re-establishes its hold across the bog.'[63]

By comparison with the raised bogs of the midlands and north, which developed from shallow lakes, the blanket bogs cover vast tracts of rocky land on the west and north-western fringes of Ireland. Often stretching from the mountain slopes right down to sea level, they tend to have a shallower depth of peat, but this is formed in exactly the same way as in raised bogs – through the partial decomposition of vegetation over thousands of years in very wet and acid conditions. In a natural landscape, the peat at lower altitudes would support a variety of native trees, but these have been gradually removed over the millennia through felling and grazing by livestock. A short walk across some peat cuttings will often reveal the stumps and roots of ancient pine trees excavated from the bog during the process of extracting turf. The wooded nature of many of the islands within bog lakes shows that, where sheep grazing is not possible, trees will persist despite the wet acidic conditions of the soil. It is heavy grazing pressure that keeps the bogs open and windswept and it is the sheep farmers who have managed this landscape over centuries. But there are some initiatives underway to address this issue.

Wild Atlantic Nature is the title of an Integrated Project funded by the EU LIFE programme and set to run for nine years up to 2029. Its focus is on damaged habitats, in particular to improve the conservation status of blanket bog, a priority habitat under the EU Habitats Directive. The primary focus is on 35 SACs spread over a huge area in the northwest of Ireland. The project works with farmers and local communities to conserve and improve the quality of blanket bogs and associated habitats, and the ecosystem services they provide include clean water, carbon storage and biodiversity. Like the groundbreaking Burren Farming for Conservation

Project in County Clare, it includes a voluntary Results-Based Agri-environment Payment Scheme. Payments to farmers are dependent on the quality of the habitat that they manage, thereby making the landowner and their skills, expertise and knowledge of the land central to the success of this project. The project will also implement a number of actions aimed at enhancing wider community engagement. These include establishing local support groups across the project sites, developing and implementing community knowledge exchange programmes, administering community outreach activities, developing a school education programme and supporting communities to develop and manage tourism and recreational activities and develop appropriate infrastructure.

The conservation of the last remaining peatlands in Ireland is really a rearguard action. In particular, so little remains of the original raised bogs in the midlands that we are dealing with tiny fragments, many of them already compromised by cutting around the edges. It is rather like trying to prevent water flowing out of a bucket when there are tiny perforations all around the sides. Some very large restoration projects have been undertaken by state or semi-state bodies such as the National Parks and Wildlife Service, Bord na Móna and Coillte. However, there are also peatland restoration projects being undertaken by non-government organisations like Ulster Wildlife, the Irish Peatland Conservation Council and local communities such as the Abbeyleix Bog Project. A contrasting approach is used for the conversion of cutaway bog where there is so little peat remaining that the original habitat is beyond restoration as peatland. In particular, Bord na Móna has identified a number of former peat production areas which are being left to nature with a combination of natural regeneration and targeted rehabilitation.

The scale and diversity of these projects is impressive and they are setting a fine example of nature restoration at scale for other European countries that have bogs. Early results are encouraging, but it is clear that this is a long-term undertaking and ongoing management is needed to ensure that the best conditions for biodiversity are maintained. Bog restoration and rehabilitation are well underway in Ireland but this will take continued commitment and finance for many decades in order to achieve the target of restoring these areas for nature and people. In the National Museum of Ireland, there is a stunning exhibition of 'bog bodies' that were buried in Irish peatlands with a selection of precious objects. Each of these objects is thought to have been associated with the inauguration of a new king and were buried at a boundary of the kingdom as a statement of the king's new sovereignty. From this it is quite clear that bogs were very important places to Iron Age and Bronze Age people. Perhaps future generations will be able to wander among vast wetlands that resemble the midland landscapes when the bogs developed many thousands of years ago.

Rediscovering the sea

On midsummer's day, I was sitting aboard a yacht in a windy west coast bay watching the sun set over the Atlantic. All around were stunning views of misty mountains, shining beaches and dark headlands, but the forecast was for storms in the next few days. With my brother, I was fulfilling an ambition to sail around the island of Ireland and, when we cast off from the east coast harbour of Greystones, County Wicklow, we had some 1,600 kilometres of sailing ahead of us before returning home. After five weeks on the sea in every kind of weather, I was left with a lasting impression of the wonder and diversity of the coast and seas surrounding Ireland.

But there is trouble at sea. The 'double whammy' of a climate crisis and a century of overfishing are decimating our seas. The coastline is also seriously impacted by coastal erosion caused by rising sea levels, pollution from fertiliser runoff and untreated wastewater from coastal towns. Although much of this environmental crisis is below sea level and thus invisible to the ordinary citizen, it is nevertheless real and immediate. There are solutions available, many well tried and tested in other countries. Charles Clover, co-founder of the UK charity Blue Marine Foundation, defines rewilding of the sea as 'Any effort by anyone to improve the health of the oceans by actively restoring their habitats and species or by leaving them alone to recover.'[64] It is a question of whether there is the political will and public support available for real measures to restore the sea.

The American writer and academic Jeffrey Bolster wrote, 'Without genuinely historical perspectives on changes in the sea we can have no idea of the magnitude of the restoration challenges we face.'[65] Few Irish people are aware that this country was once the base for a Norwegian whaling operation that killed and butchered hundreds of great whales from the Atlantic Ocean. I have been to visit the site of this whalemeat factory on the island of South Inishkea off County Mayo. Here, on the little promontory of Rusheen, lie the remains of an industrial operation based on the giants of the ocean. The brick walls and rusting machinery of the whaling station lie dismembered in the ocean winds more than a century after they were abandoned. A second station was built on the mainland of Mayo at Blacksod Bay, but little trace remains of this apart from a scattering of bricks. Whaling boats from these locations killed and brought back to Ireland a total of 683 whales of six species over a seven-year period from 1908 to 1914. The most numerous quarry species comprised fin whales, but it is hard to believe that they managed to catch 66 blue whales, rarely seen in Irish waters today. The outbreak of war in 1914 put an end to the whaling operation here, although the station at Blacksod was briefly used in the 1920s, catching a further 202 whales in these years. It all ended when the Blacksod whaling station was destroyed by fire in 1923, but the demise of this local industry was probably hastened by the replacement of whale oil with fossil fuels that still largely power our economy today.[66]

Commercial whaling was banned in 1986 by the International Whaling Commission after some species almost became extinct. Norwegian, Icelandic and Japanese vessels still hunt whales today, but most other countries have banned this barbaric activity. Although it is difficult to estimate population sizes in these animals due to their long migrations, there are signs of recovery in some species. In the 20th century, sightings of the distinctive humpback whale in Irish waters were rare. Since 1999, the Irish Whale and Dolphin Group (IWDG) has been collecting humpback whale data in Ireland, including sightings by ordinary citizens. These stunning creatures, which often show their distinctively marked tail flukes when they dive, are most often seen around the south and south-west coasts. With ready availability of high-quality cameras, at least 116 individual humpback whales have now been identified from their markings. The repeat matching of photos has confirmed that many of these individual whales visit Irish waters annually. This has allowed reliable estimates of whale abundance to be made. It is now estimated that there is a superpopulation size of around 145 humpback whales with numbers increasing in inshore Irish waters, especially since 2015.[67] Many of these whales return year after year with an average resighting rate of around 63 animals in years when ten or more individual whales were recorded.[68] Dr Simon Berrow of the IWDG says,

> Clearly humpbacks are increasing due largely to a stop to killing them and their distribution might also have been affected by climate change. Fin and sperm whale populations may possibly be recovering too but not the blue or right whales. Gray whales were extinct in the Atlantic but are now recolonising as the population expanded through the opening up of the north-west passage from the Pacific due to climate warming. Other than that, none of the other cetaceans shows any population increase or recovery in Ireland.

As I sailed around Cape Clear Island, the most southerly point of the Irish coast, the glassy surface swell was broken by a series of fins. A small group of dolphins surfaced alongside me, playing in the bow wave, leaping and racing ahead of the boat. As I sailed away, I saw them repeatedly diving in the same area. They had found a shoal of sprat trapped in the sound between the islands and they were working together as a team to herd their prey into a ball. All around the boat, there were gulls, shearwaters and kittiwakes picking up the fish scraps left on the surface. While most of Ireland's seabird populations are in a reasonably healthy state, kittiwake numbers have declined by over one-third since the 1980s and breeding success has fallen markedly.[69] Poor breeding performance has been linked to declines in food availability, due to a booming herring and sprat fishery in the Celtic Sea. The practice of pair-trawling of spawning inshore sprat and herring has increased in recent years and there are likely to be implications for the breeding success of kittiwakes and dolphins along these coasts.[70] The current herring fishery in Irish waters is a small fraction of the historic catch because this species has been consistently overfished. Currently, there is no total allowable catch for sprat and this may be impairing recovery of the stock. Sprat is short-lived and an important prey fish for many marine species including whales and dolphins.

The seas and shores around Ireland are a vitally important part of the environment of this small island, but they have often been overlooked by a nation that has traditionally been focussed on farming the land. The result is that the sea has been overexploited for natural resources such as fish and shellfish and is often viewed as merely a place to dump our waste. When I was young, there were open tipheads on the beaches in Dublin Bay. Most of the sewage collected in coastal towns was simply discharged without treatment to the sea. People thought that the oceans were vast enough to absorb any waste we threw into them. A century ago, it was believed that the fish in the sea were so abundant that they would last forever.

Today, according to the Irish Marine Institute, only 20 per cent of the commercial fish stocks in Irish waters are in good environmental status. Bottom trawling is not only removing far more fish than can be replaced naturally, but it also decimates the marine communities that inhabit the sea floor. What if there was a prohibition on all fishing in certain marine protected areas? How would the coastal communities who have come to depend on this natural resource survive such measures? They are already well used to adapting to bans on the catch of certain fish species when the stocks have reduced to critical levels, so perhaps they would adapt to this too.

Take, for instance, the Atlantic salmon. By 2006, salmon catches in Ireland and elsewhere had seriously declined in the years since an historic high in the mid-1970s. Numbers of fish returning from the sea to rivers had declined by more than two-thirds and were the lowest on record for the previous 35 years. The fact that salmon stocks in many countries bordering the North Atlantic were affected suggested that a wide range of factors were contributing to the decline. Overfishing at sea was clearly a major cause as the total catch by drift-netting had fallen sharply to about half of what it was in 2001. Two-thirds of the salmon drift-net fishermen in Ireland were catching less than 100 salmon each year. The report of the Independent Salmon Group, established by the government, made a number of key recommendations to tackle this problem. The report warned that radical measures were necessary to halt 'the catastrophic decline of Irish salmon stocks'.[71]

The result in 2007 was a complete ban on drift-netting for salmon at sea and a ban on angling for salmon in a number of vulnerable rivers in the east and southeast of the country. Those involved in drift-net fishing were compensated for the complete closure of the fishery. Inshore commercial fishermen were offered €25 million in compensation to quit, though not all licence holders participated.

However, ten years after Ireland banned drift-netting for salmon at sea, the controversial measure had not changed this steep decline. The Atlantic wild salmon was in a 'serious position' and various conservation measures to save the salmon 'don't seem to have worked', according to Dr Ciaran Byrne of Inland Fisheries Ireland. Speaking at a conference in Galway, Dr Byrne said marine survival of the species had fallen from 20 per cent of fish returning to rivers in 1980 to five per cent in 2017. Dr Ken Whelan, fisheries scientist for the EU Salsea project, found that climate change was contributing to an 'alarming' increase in the numbers of wild salmon dying at sea, and that the stock was moving north in response to a warming ocean. The project found salmon were feeding at the very edge of frozen polar ice fields – sometimes at depths of up to 800 metres, normally inhabited by the sperm whale. Dr Whelan warned that Ireland was among a group of 'southern stock states' where the wild salmon was threatened with extinction if high mortality at sea continued.[72]

Walking on the beaches near Arklow in County Wicklow I never fail to marvel at the huge quantities of empty native oyster shells on the strandline, but this species is effectively extinct in this part of the east coast, so these shells are very old. Centuries ago, the flat oyster was an abundant shellfish in European seas but sadly it is estimated that populations have declined by 95 per cent since the 19th century. Now native oyster reefs are one of the most threatened habitats in Europe. At one time, almost every bay and many offshore banks around the Irish coast had natural beds of oysters that were thought at the time to be inexhaustible. The exploitation of this wonderful natural resource was virtually unregulated except by taxes levied by both church and civil authorities. Some beds, such as those in Dublin Bay, were rented out by Dublin Corporation, which required that the lessees restock the beds with oysters dredged from elsewhere. In Carlingford Lough, there were over 400 small boats engaged in dredging the public oyster beds. Huge quantities of this much-valued seafood were exported to England and France to feed the masses of the industrial revolution. But like all overfishing practices, it brought about its own downfall. By the mid-19th century, many of the beds were exhausted and attempts were made to cultivate them in artificial 'oyster parcs'. This met with limited success and was followed by the introduction of the Pacific oyster to many bays.[73]

In Northern Ireland, extensive oyster beds are known to have existed in Carlingford Lough, Lough Foyle and Belfast Lough for several hundred years, but most stocks crashed during the late 19th or early 20th century. Now only a small wild population exists in Strangford Lough.

In 2020, live oysters were discovered in Belfast Lough for the first time in over a century, providing evidence that the environmental conditions for reestablishment had, at least partially, returned. To support the natural recovery of the native oyster in Belfast Lough, the non-governmental organisation Ulster Wildlife has installed dozens of oyster nurseries in several yacht marinas, including Bangor, County Down, in the first project of its kind in Northern Ireland. An oyster nursery is a micro-habitat housing mature oysters that will reproduce and release the next generation of spat (or larvae) to settle out on the seabed of Belfast Lough. An individual mature oyster can release up to one million spat per year and these will settle on suitable habitats nearby. The oysters are in cages hung in the water beneath the pontoons, providing them with protection from predation by birds.

I lay down on one of the pontoons in the small marina at Glenarm, County Antrim, for a closer look at the oyster cages. Coated with weed, they looked like something that had been recovered from a wrecked ship, but these cages hold the key to restoring a marine ecosystem that once thrived right around Ireland. I spoke to the Marina Manager, Billy McClelland, a former fisherman who remembers native oysters being dredged from the banks off the East Antrim coast. He has given his wholehearted backing to this restoration project. In partnership with the University of Bangor, Ulster Wildlife has conducted intertidal surveys of Belfast Lough and Strangford Lough and they plan to survey Carlingford and Larne loughs to assess the status of the remaining native oyster population. The results will be used to assess the feasibility of large-scale restoration of oysters in Northern Irish waters.

A similar project is already underway in Dublin, run by University College Dublin in collaboration with Watermark Coffee and their Green Ocean initiative. This involves the sourcing of juvenile oysters from Tralee Bay, where they are free of the disease *Bonamia*, and placing these in specially designed boxes hung beneath the pontoons in three marinas at Dún Laoghaire, Poolbeg and Malahide. The oysters typically grow to maturity in three years after which they will spawn, and the spat will hopefully settle in the vicinity of the marinas. After a few years, it is planned to place the mature oysters in suitable locations on the seabed with the objective of restoring a viable native oyster broodstock in Dublin Bay and surrounding areas. These new oyster beds will be carefully monitored by the UCD team to measure the success rate. The boxes in the marinas will then be replenished with new juvenile oysters. The native oyster is a keystone species with many benefits for the marine environment. When oysters are restored to the sea, their ecosystem functions are re-established, helping to improve water quality and clarity. Each day a single oyster can filter up to 200 litres of seawater (enough to fill a bathtub), removing particles from the water column. Left undisturbed, oysters will form reefs, restoring natural habitats for many marine species including important commercial fish. The oyster can also increase light penetration to the sediment and promote the recovery of seagrasses, another threatened and valuable coastal habitat.

Some years ago, with colleagues, I undertook an extensive survey of seagrasses in the estuaries of Ireland to find out what remained of these threatened habitats. Walking across the wide mudflats in places like Lough Foyle, County Derry or Castlemaine Harbour in County Kerry brought feelings of a wilderness surviving close to human settlement. This was especially so when we encountered the meadows of green grass-like plants that grow near the low tide mark. These are eelgrass or *Zostera* beds, which are like fast-food takeaways for brent geese and other wildfowl when they first appear each autumn to overwinter in Ireland. They offer a high-protein food that the birds need to replace the body weight lost during their migrations. In many bays around the west coast, there are also subtidal eelgrass beds that grow permanently in shallow water. When snorkelling in Mannin Bay, County Galway, I have seen these extensive seagrass beds with their leaves waving back and forth in the current. I sailed over similar seagrass beds growing on very shallow sandy habitats in Blacksod Bay, County Mayo. Like all green plants, they depend on sunlight to photosynthesise and grow, with clear water and shallow seas providing exactly the right conditions. Healthy seagrass beds support a wide diversity of animal life both above and below the seafloor. They offer safe nursery areas for fish like cod and herring and shellfish such as scallops. Sharks and rays also use the cover of these marine meadows to lie in wait for their prey. Globally, at least 50 species of fish live in or visit seagrass beds and about one-fifth of the world's biggest fisheries are supported by seagrass meadows as fish nurseries. Importantly, in the current climate crisis, seagrass meadows store carbon as effectively as forests at about 400 kilograms of carbon dioxide per hectare each year. This makes seagrass meadows as important for climate stabilisation as native woodland on land.

Seagrass communities are declining across the world's oceans and Ireland is no exception to this. Globally, estimates suggest that we are losing an area of seagrass around the size of a football pitch every 30 minutes. One of the main causes of this decline is pollution draining into the sea via rivers from agriculture, forestry and urban settlements on land. Nutrient enrichment through the discharge of phosphates and nitrates causes other more vigorous plants to grow on top of the seagrasses, blanketing them from the vital sunlight. Declines of this sort have been noted in bays all along the west coast from Castlemaine Harbour in County Kerry to Mulroy Bay in County Donegal. Scallop dredging also causes serious damage to seagrass beds, such as those in Blacksod Bay, according to reports by the Marine Institute and the NPWS.

Similar declines have been recorded in Britain, but there, various organisations are taking action to restore the seagrasses. Seeds are collected from the existing eelgrasses when they mature and these are grown in laboratory conditions so that the plants can be later sown back out on the ocean floor. In Wales, Seagrass Ocean Rescue is undertaking a major restoration project planting five million seagrass seeds off the Llŷn Peninsula in Gwynedd and Anglesey, with the aim of creating ten hectares of seagrass meadow by the end of 2026. It is estimated that a single hectare of seagrass could harbour up to 80,000 fish and 100 million invertebrates. Launched in 2019, England's largest seagrass restoration project had planted around 70,000 seed bags by 2022, spanning more than three hectares of seabed in Plymouth Sound and the Solent. This will provide vital habitat for marine life. In Scotland too, a large restoration project has been started. The Restoration Forth project has successfully injected 25,000 seeds into the seabed at three trial sites in the Firth of Forth. These and many other similar projects worldwide have trialled and undertaken practical restoration work, so the methods are well proven.

I sailed recently in Tralee Bay in County Kerry and was especially struck by the huge areas of shallow water over a sandy seabed. Dr Liam Morrison of the University of Galway is leading a seagrass restoration project here. His team have been snorkelling in the bay, collecting samples of seagrasses for genetic analysis with a plan for transplanting these to other areas where seagrasses have declined. So far, they have transplanted eelgrass in both intertidal and subtidal habitats, using shoots and seeds, in Kilkieran Bay, Connemara. Some shoots were also transplanted from Kilkieran Bay to Rusheen Bay near Galway City. Monitoring the success of these transplants and the effect of winter storms will guide future restoration work in these areas. There is no reason, other than finance, why similar seagrass restoration projects could not be undertaken in many bays on Ireland's coast, as these shallow areas are largely unused by any commercial interests other than oyster cultivation in licenced areas. However, the source of pollution entering the bays needs to be meaningfully addressed at the same time.

Many times, I have also snorkelled amongst beds of brown seaweed on the Irish coast. Waving back and forth in the tide, they are like forests that shelter whole communities of fish and other marine animals. Mammals such as seals and otters often use these kelp beds for shelter and for feeding. Typical invertebrate creatures include the edible sea urchin, common starfish, brittlestars, blue-rayed limpet, topshells and paddle worms. I once saw a dogfish 'asleep' on the sand among the moving seaweed. The long, oar-like blades of kelp are colonised by encrusting animals like marine worms and sponges while other animals feed on these algae. They are attached to the seabed by strong holdfasts that look like surface roots. At extreme low tides, the kelp blades can be seen on the surface, but they don't stay in the air for long. After a storm, the holdfasts may loosen and the large fronds of weed are washed onto beaches where they provide food for other shoreline creatures. In the past, the rotting seaweed was a valuable resource for coastal farmers who harvested it and spread it on the land to add nutrients to poor acid soils.

A survey of the seaweed resources off the west coast of Ireland in 1998 found that kelp beds were common or abundant at nearly half of the sites surveyed with the dominant species being oarweed and the cuvie.[74] The oarweed has a flexible stipe (or stalk) which holds the kelp upright when covered by the tide but allows it to bend over under the pressure of wave action. In this way, the oarweed can live at a higher level on the shore than the cuvie. At Spiddal, County Galway, the kelp beds cover a 100-metre-wide belt parallel to the low tide mark. With densities of 10 to 15 plants per square metre, the kelp beds resemble a forest on land. Mature plants form the canopy while young saplings spring from below. Just as in trees, the age of the plants can be calculated by counting the growth rings on a cross-section through the stipe, with the average age of mature plants estimated at just five years. In the age of rapid climate change, it is known that a kelp forest can take up to 20 times more carbon dioxide from the atmosphere than land-based forests.[75]

Unfortunately, kelp beds in deeper water can be destroyed by inshore trawling. On the south English coast, a pioneering marine project has been launched to restore almost 200 square kilometres (about the size of Liverpool) of lost kelp forest along the coast of Sussex. The project is focussed on four species of subtidal seaweeds – oarweed, tangle or cuvie, sugar kelp and furbellows. Historically, kelp was abundant along the Sussex coastline, but this important habitat has diminished over time, leaving just a few small patches and individual plants, mostly in shallow water and along the shoreline. The Sussex Kelp Recovery Project began after the implementation of the Sussex Nearshore Trawling Byelaw in March 2021, which ensured that the nearshore seabed off the Sussex coast was protected from bottom-towed trawling gears. The natural recovery is being monitored closely by marine scientists using video cameras. Active restoration initiatives like planting or seeding kelp can also help to kick-start the process.[76] The project is led by Sussex Wildlife Trust and a range of other supporters. Their vision is to see the recovery of kelp and other essential fish habitats at scale in Sussex, supporting a thriving and sustainable marine ecosystem that benefits nature, fisheries, coastal communities and the planet. Similar projects could easily be implemented in Marine Protection Areas (MPAs) in Ireland with the collaboration of coastal fishing communities.

I was sailing off the coast of Mayo recently when a pair of fins appeared in the water beside the boat, moving slowly on the surface. My first impression was that these were dolphins, but there was no diving or breaching involved. As the fins came closer, I realised that they were both part of the same animal: a basking shark, the largest fish in the Atlantic Ocean. It moved sinuously across the sea with a large open mouth capturing the rich harvest of plankton. Basking sharks were once plentiful around the Irish coastline, migrating into our waters in summer and disappearing in the winter. In previous centuries, Achill Island, County Mayo was one of the locations for the systematic capture and killing of large numbers of basking sharks mainly for their liver oil, which was used for lighting lamps in the days before electrification. A shortage of fuel after World War II once again led to an increasing market for shark oil for use in certain industrial products. The slow-moving sharks swam into the bays to feed on dense swarms of plankton near the sea surface. There they became entangled in nets set by the islanders who then launched their lightweight curraghs and killed the struggling fish, stabbing them with scythe blades attached to long poles. Over the thirty-year period up to the 1970s, more than 12,000 basking sharks were landed on Achill – an average of at least 400 fish per year. These harmless sharks were caught at a number of regular haunts along the west coast in the preceding century and large quantities of the valuable oil were exported to England. Not surprisingly, catches declined markedly towards the end of this period and, with the availability of alternative mineral oils, the market for shark oil disappeared, allowing the few remaining animals to survive. [77]

Today, the basking shark population is recovering from centuries of overfishing. Malin Head, County Donegal, the most northerly point of Ireland, is one of the best places to watch these enormous fish in summer. Emmet Johnston of the NPWS has estimated that up to a quarter of the world's basking shark population moves through the sea off the north coast of Ireland at certain times of the year. As a result of research and publicity by the Irish Basking Shark Group (IBSG), Irish waters are now internationally recognised as providing vital habitat for large numbers of these sharks feeding along our coasts during summer and courting during autumn months. After a century of decline, it took until 2022 for the basking shark to be legally protected under Ireland's Wildlife Act. Dr Simon Berrow of the IBSG said,

> Unfortunately, during the summer, unintentional harassment and disturbance of basking sharks is a growing problem. Boats are asked to avoid them while the sharks are feeding and courting in Irish waters. This is why legal protection and the proposed code of conduct are so necessary because they make the disturbance and harassment of basking sharks illegal from now on.

One of my favourite ways of relaxing while out sailing is to throw a line, with some white feathers on the hooks, over the side of the boat in the hope of catching one or two mackerel for my dinner. I can remember a time when I would catch up to six beautiful green and black fish with one cast. However, the frequency and size of my catches has been declining each year and this is reflecting changes in the stocks of these pelagic shoaling fish. They are being hoovered up by giant trawlers all across the North Atlantic from Spain to Norway. I saw some of these trawlers, each as large as a football pitch, tied up to a pier in Killybegs, County Donegal. The scientific limits to the catch are set by the International Council for Exploration of the Sea (ICES) and are based on the most recent population estimates. While most countries agree on the scientific limits to the catch, they do not agree on how to share it, meaning that overall catches usually exceed the advised amount. There are 17 mackerel-fishing countries in Europe (including Ireland and the UK) and together they have overstepped the ICES threshold by an average of 41 per cent each year since 2010. Mackerel numbers have been declining since 2015 partly due to fewer young fish surviving to adulthood. As well as overfishing, this could be due to changing ocean temperatures, less available prey and disease. In 2023, these 17 countries could not agree how to divide the mackerel catch fairly between them and so the overfishing continues. In response, the Marine Conservation Society has increased its rating for mackerel to 'amber' as a way of drawing the attention of consumers to the threat to the species. This is just the latest in a long history of mismanagement of commercial fisheries in Europe.

A few years ago, a number of non-government organisations got together and set up the joint campaign Fair Seas. The campaign seeks to protect, conserve and restore Ireland's unique marine environment. Its ambition is to see Ireland become a world leader in marine protection, giving our species, habitats and coastal communities the opportunity to thrive. This is a fairly ambitious target given the centuries of mismanagement of our marine resources. So too is the target of having 30 per cent of Ireland's maritime territory strictly protected in MPAs by 2030, when the current legally protected areas (designated under EU law) cover just over 8 per cent of our waters. Designating such sites at present simply involves drawing lines on maps and setting conservation objectives for them. It does not guarantee that they will be managed any better.

The need to expand the network of MPAs in Irish waters is recognised by the Irish government in order to address the current situation which is not satisfactory. An expert advisory group has concluded that, 'Ireland's existing network of protected areas cannot be considered coherent, representative, connected or resilient or to be meeting Ireland's international commitments and legal obligations. There is no definition of MPA in Irish law and this is a gap which needs to be addressed.'[78] Encouragingly, the Irish government has acted by beginning to draft new MPA legislation. Having this national legislation will be key to how well we can designate and manage our protected areas for biodiversity and deliver the greatest benefits to our coastal communities. How such 'protected areas' are managed remains to be seen. Where fisheries, aquaculture, dredging or dumping at sea are considered to be detrimental to the natural balance in an MPA, then these activities should be suspended or should be banned altogether until the ecosystem recovers. Conservation objectives and management plans for MPAs need to take account of the historical richness of our seas instead of setting the benchmarks at maintaining a status quo. The plight of the native oyster, which was driven to extinction in most of our inshore waters some centuries ago, is a perfect example. Restoration of this key mollusc to its former habitats would have enormous benefits for the marine ecosystem. If all Irish coastal waters were managed sustainably, as they ought to be, this would, over the long term, benefit both fisheries and marine conservation and eliminate the need for MPAs.

In scaling up Ireland's MPA network, the Fair Seas campaign says it is clear that a more ambitious approach is urgently needed to ensure the health of marine biodiversity is restored. The most significant pressure on marine biodiversity in Ireland is the overfishing of fish and shellfish and disturbance associated with the fishing activity. Fair Seas is therefore calling for fully protected marine areas which do not allow any kind of extraction through fisheries or aquaculture, while highly protected areas allow only small-scale, low-impact extraction of fish or shellfish. Neither protected area category would allow dredging or dumping, and only minimal- to low-impact, small-scale infrastructure would be permitted. This would also rule out industrial-scale fishing and all fishing using trawling gears that are dragged or towed across the seafloor or through the water column, and fishing using purse seines and large longlines. A simple measure, that would aid fishing vessels in recognising and respecting protected areas, would be to show the boundaries of MPAs on the marine charts that are widely used both in paper and digital formats. Most modern boats use digital charts with satellite navigation so it should be a relatively easy process to update these.

Globally it is well established that fully protected areas will eventually hold a much higher density and diversity of marine life as well as more large individuals. As the animals within a fully protected area grow larger over time, they also produce more eggs, are more successful at reproduction and produce fitter young. At present, there is no fully functioning Marine Protection Area in Ireland that strictly conserves fish stocks, so there is nothing that can act as a template. Accepting that there are designated marine Special Areas of Conservation and Special Protection Areas (for birds), these carry no prohibition on overfishing, and it is quite likely that most of the fishing fleets, especially overseas vessels, are completely unaware that such designated areas even exist. In 2023, the Minister for Agriculture, Food and the Marine signed off on fisheries management measures for a small part of Dundalk Bay SAC. The Fisheries Natura Declarations set out a number of prohibitions and restrictions in a small area inside the bay which is mapped out on the Sea-Fisheries Protection Authority website. The first declaration prohibits fishing with specified types of fishing gear including dredges, beam trawls and bottom otter trawls. It also sets out monitoring requirements for certain sea-fishing boats operating in Dundalk Bay, which is inside Ireland's Natura 2000 network. The second declaration sets restrictions on fishing, dredging and hand gathering in the bay. It is a small start, and it needs to be scaled up and extended to many other areas that are overfished. How these new restrictions will be enforced has not been defined and there is no provision to monitor the results.

A good example of a working MPA is provided by the Lyme Bay Fisheries and Conservation Reserve, a protected area in southern England in which multiple uses including fishing are allowed as long as none are damaging to the seabed or to nature conservation. The project has forged links between fishermen, conservationists, regulators and scientists in order to maintain a healthy, productive and sustainable Marine Reserve within the bay, that will benefit fishermen and conservationists alike. The Reserve has achieved three objectives: to protect the biodiversity of Lyme Bay, to implement best practice in managing fish and shellfish stocks and to create long-term benefits for coastal communities around the bay. One of the aims here is to help fishermen achieve best quality and top pricing for their catch. To do this they have created the 'Reserve Seafood' brand, which markets the low-impact, sustainable, premium quality, provenance-assured seafood of Lyme Bay. There is clearly a premium market for sustainable seafood, so each fisherman is signed up to the Lyme Bay Fisheries and Conservation Reserve and is accredited by the Seafish Responsible Fishing Scheme. This scheme assures catch quality and best fishing standards. The voluntary Codes of Conduct that each fisherman adheres to and the science which underpins the results of fishing efforts in the Bay inform the sustainability of the product. Each vessel is also fitted with an inshore Vessel Monitoring System which guarantees the low-impact traceability of each catch. All of this helps towards the long-term sustainable future of Lyme Bay for both fish and fishermen.

In Scotland, the area known as Lamlash Bay, on the east side of the Isle of Arran, has been declared a no-take zone for the last two decades. The Firth of Clyde was once known for plentiful fishing of herring, cod, haddock and turbot. Small-scale traditional fishing provided a livelihood for generations of families in the Clyde and Arran, who were able to fish sustainably thanks to long-standing laws banning practices that towed fishing gear along the seabed. The decline in these fisheries was caused by the 1984 removal of the ban on bottom trawling within three miles of the UK coast. The resulting decimation of the fish stocks had an evident impact on Arran's community. In the 1994 international sea-angling festival, catches dropped by 96 per cent. In 2008, the first community-led No-Take Zone (NTZ) was established in Lamlash Bay. In this small area of less than three square kilometres, no fishing of any sort is permitted. In 2016, the South Arran Marine Protected Area extended this area to nearly 300 square kilometres to exclude scallop dredging but allow for other, potentially more sustainable, fishing methods in various zones.

In 2010, the University of York initiated a project in collaboration with the organisation COAST to study the differences between the seabed protected within the NTZ and other areas. To do this they used many methods: diving surveys (counting what they saw), drop-down baited cameras and lobster and crab surveys with local fishermen. They found that biodiversity had increased by 50 per cent and the populations of scallops and European lobsters were 2–3 times higher within the NTZ. There were growing numbers of larger lobsters and evidence of lobster 'spillover' into surrounding areas. The results also reveal that there are nearly four times more king scallops in the area since 2010, the size of adult scallops has increased and there are more juvenile scallops present in the area. Additionally, dive surveys of the area show the seabed is recovering after damage caused by fishing with trawls and dredges, with the growth of structurally complex 'nursery habitats' which provide refuge for marine life. The findings are acting as a blueprint to support similar areas and other coastal communities around Scotland and further afield and underline the case for increased marine protection.[79]

In Northern Ireland, new fisheries management measures were enacted in 2023 which will prohibit bottom trawling in nine inshore MPAs. There will also be measures put in place in these MPAs for static gear such as pot fishing. This is in stark contrast to the reversal of a ban on boats longer than 18 metres trawling inshore waters in the Republic. The Northern Ireland Marine Task Force, along with other stakeholders including members of the fishing industry, have formed a new coalition called Co-Fish: NI Fisheries and Conservation Partnership. They will work on ensuring that these management measures will deliver benefits for the marine environment through restoration and conservation while ensuring livelihoods of the fishing industry are also protected. They also plan to ensure communication between government departments and stakeholders so their MPA network can be improved and managed as effectively as possible.

Ensuring engagement from stakeholders such as the fishing community will be vital in the process of managing a MPA once it is declared. Typically, the use of the sea is centred around the seafood sector which currently employs about 16,000 people in Ireland. According to the Bord Iascaigh Mhara Seafood Report for 2020, the total value of Ireland's seafood economy was just under €1.1 billion. The World of Organic Agriculture 2021 report declares that Ireland had the largest production of aquaculture products (mainly shellfish) in Europe during 2019, with more than 27,000 metric tons. However, the problem of overfishing has not gone away.

Damage to marine biodiversity is far from the thoughts of most Irish people. Hidden beneath the waves for the most part, it is out of sight and out of mind. Overexploitation of marine resources has been going on for centuries and, with the addition of warming seas and acidification due to climate change, it is now at a critical point. As I sailed away from the basking shark off the Mayo coast, leaving it alone to continue feeding, I was full of hope that these impressive fish are now recovering from centuries of overfishing. There are plenty of examples from other countries which show that, where the pressures are reduced, the sea has a capacity for rapidly restoring itself. It just needs a little help to kick-start the process. Charles Clover wrote: 'To do so we just need to trust in nature a little more and invest in the natural world in the only way we can, by being a little more prepared to leave it alone.'[80]

Losers and winners

Keen nature-lovers might be understandably confused when they read on the one hand of the loss and endangerment of many Irish native species, like breeding curlews and wild bees, or the decimation of natural habitats like peatlands and woodlands, and then another account of the amazing return of the buzzard, pine marten or the great whales. We could call these the losers and winners. The key to understanding what is happening lies in the fact that species and ecosystems differ in their resilience to change, whether this is caused by human or natural agencies. Climate change is now also a driver of dramatic alterations in the range of some species.

Our knowledge of how species arrived in Ireland in the first place depends largely on archaeology. Just after the start of the twentieth century, a group of early archaeologists began the excavation of two limestone caves near Ennis, County Clare. Over a period of years, they amassed a huge collection of over 50,000 bones which was donated to the National Museum of Ireland.

More than a century later, two research scientists, Dr Ruth Carden and Dr Marion Dowd, began re-examining these bones and among them they made a startling discovery. This was the kneecap of a brown bear on which there were unmistakable cut-marks of a tool suggesting butchering by humans. However, the most remarkable fact emerged when the bone was radiocarbon-dated to around 12,500 to 12,800 years ago. At that time, most of the northern part of the globe was still in the grip of a significant climatic cold period marking the end of the last ice age and Ireland was still covered with a tundra landscape. It would have been an unwelcoming place for early human settlers. And yet this single butchered bone from County Clare now suggests that there was hunting of wildlife in Ireland much earlier than previously thought.[81] Ruth Carden says, 'Up to now we have not factored in a possible "human dimension" when we are studying patterns of colonisation and local extinctions of species to Ireland.'

The ancient hunter-gatherers would have encountered a very different landscape to that which we see in Ireland today. Farming was unknown in northern Europe and the vegetation was entirely natural. By the time there were established settlements, a number of the larger animals of the earlier periods were already gone from Ireland. During the warmer periods between the advances and retreats of the glaciers, giant Irish deer, reindeer, wolf, arctic fox, lemming and brown bear had roamed the Irish landscape, but the severe cold eventually led to their extinction. Stoat, mountain hare and red deer are thought to have been here too in the earlier period. However, the Mesolithic hunter-gatherers, who arrived when the glaciers finally receded, shared the landscape with brown bears, wild boars, wild cats and lynx, some of which may have been introduced.[82] Remains of brown bears have been found in various parts of Ireland – Ailwee Cave in County Clare and Killaun Bog in County Offaly are two examples – but their preservation does not confirm that they lived in caves, just that their bodies ended up there.

According to Dr Nigel Monaghan, formerly curator of the Natural History Museum in Dublin, brown bear bones have been found in at least 32 locations in Ireland and these have been dated to two phases of prehistory, 45,000 to 30,000 years ago during an interglacial warm period and 12,000 to 3,000 years ago when they definitely overlapped with early humans. It is likely that they lived in the extensive forests that colonised Ireland in the post-glacial period. An alternative suggestion is that the bear was reintroduced in the Mesolithic as this species was known to have been kept in captivity in Europe at this time, possibly for symbolic reasons.[83] A single femur from a lynx was found in Kilgreany Cave in County Waterford in the 1920s and this was recently dated to around 10,220–9,945 years before the present. This is the only known archaeological evidence that lynx ever lived in Ireland.[84] Wild cats and lynx would certainly have been valued in Ireland for their fur, and boars would have been a very useful source of meat at a time when livestock were unknown. Some of the other early mammal species managed to hold on much longer. The wolf survived in the remaining fragments of woodland with the last of these iconic animals being killed in 1786.[85]

Hunting for food, combined with the loss of its woodland habitat, was almost certainly the cause of the early demise of the capercaillie, a large, turkey-like bird that is known in Ireland only from the bones found in archaeological sites. It appeared first in the earliest known human settlement at Mount Sandel in County Derry and again in many sites right up to medieval times when it disappeared altogether. Likewise, the crane was a common enough bird in ancient times. Large and slow-flying birds that spent a lot of time on the ground, they were relatively easy to hunt and were a popular food item with ancient people. The last period in which crane bones have been found was the 17th century in Roscrea Castle, County Tipperary. These two species still occur in other parts of their range abroad. But a third, the great auk, which is known to have bred in Ireland from prehistoric times until the early 19th century, is now extinct throughout the world, just like the dodo. This bigger relative of the puffin was sometimes described as the 'penguin of the north' as it was flightless and almost a metre in length. It would have been relatively easy for coastal people to hunt these birds when they came to land for nesting.

Birds of prey were certainly not hunted for food, but they were persecuted for a variety of other reasons including sport, egg-collecting and game protection.

Healthy populations of white-tailed and golden eagles, osprey, marsh harrier, goshawk, buzzard and red kite all disappeared with the final extinction for most occurring around the end of the 19th century.[86] There were certainly innumerable other smaller animals, such as insects, molluscs, fish and many plants, that also became extinct in this period due to habitat loss, such as the virtual clearance of the ancient woods by 1600. These went largely undocumented, as reliable natural history recording did not begin before the 19th century.

Most of the plants that are known to have become extinct were already quite rare in Ireland when first recorded, usually because they grew in unusual habitats. Buxbaum's sedge grew on an island at the north end of Lough Neagh, where it was first found in 1835, and survived at least until 1886. Scrub encroachment, heavy grazing and lowering of the water table have all been suggested as causes of decline. Descendants of the plant survive in cultivation. Rannoch rush disappeared within a few years of its discovery on a bog in County Offaly owing to peat exploitation. Saw-wort was known from a rocky riverbank in County Wexford, where it survived until 1952, but has not been seen since. Five other species believed to be extinct include purple spurge, scaly buckler-fern, holy-grass, sea knotgrass and fen woodrush. Oak fern might also be extinct. It had been seen on widely separated Irish mountains over a hundred-year period but was last recorded in County Antrim in 1979. The rapid changes in farming practices in the 20th century led to the loss of many species. The last Irish records of meadow saxifrage as a native were from counties Wicklow in 1985 and Dublin in 1986. Historically it grew in moist but well-drained, often lightly grazed, calcareous and neutral grassland as well as in unimproved pastures and hay meadows. Despite repeated searches of these sites specifically for the species it has not been refound. It was officially declared extinct in the Red List of Irish Plants in 2016.[87]

There were originally 596 different species of moss recorded in Ireland, but thirty-five of these are now extinct here. For example, the spiral chalk moss was recorded in and around Dublin city during the mid-19th century but has not been recorded in Ireland since 1870. It grew on mud-capped stone walls, a habitat that has now disappeared. Some insects have also disappeared, but these were largely undocumented due to the lack of experts in the past. The mountain ringlet was a small brown butterfly known in counties Mayo and Sligo during the 19th century, but it has not been seen since 1901. It was found on the higher slopes of mountains and reputedly only flew in bright sunshine. Today there are only a handful of specimens in museum collections. The hornet moth is a large day-flying moth that mimics a real hornet's jerky flight style. Its larvae feed on tree tissue just under the bark. The species was last sighted in Ireland in 1946, but the causes of its apparent extinction are unknown.[88]

So, a picture emerges: significant numbers of larger animals and plants, and uncountable small or poorly studied species, vanishing from Ireland with the pressures of habitat loss, hunting and persecution over the millennia since humans arrived in this country. Of course, many others hung on, but changed from being common inhabitants of the Irish countryside to rarities confined to a few isolated fragments of habitat. For example, in the 19th and early 20th centuries, the corncrake was a common breeding bird in hay meadows and other habitats throughout the country, including the outskirts of Dublin city. With the growth of mechanised farming, it declined dramatically, and there are only a handful of breeding pairs today, restricted to some tiny islands off the west coast. Likewise, freshwater pearl mussels were once so common in rivers throughout Ireland that they were harvested by country people searching for the pearls they contained.[89] Due to a decline in water quality across Ireland, the majority of the mussel populations have failed to reproduce in recent decades. Today there are only a handful of rivers with healthy populations of this large mollusc and the freshwater pearl mussel is very likely to become functionally extinct in the wild in the near future. Corn marigold is found in farm crops and occasionally on other disturbed ground such as roadsides and waste ground. It was probably introduced in Neolithic times and was formerly a weed of spring-sown cereal, but is now much reduced by improved husbandry, seed cleaning and herbicides. Its showy, golden-yellow flowers make it a favourite ingredient of wildflower seed mixtures. But as yet, this does not appear to have arrested the decline of the species as a wild Irish plant. It is now uncommon generally, except in arable areas near the coast, notably County Wexford.[90]

Nobody knows exactly how many species have been lost from Ireland since the last ice age, but the missing animals and plants are almost all specialists that require quite particular habitat conditions and will not return easily unless these habitats are available. Similarly, where persecution, poisoning or hunting have caused the losses, there is no question of a recovery without a reduction in these activities. However, some larger species do give cause for hope, having returned of their own accord. The common buzzard was indeed a common species in Ireland before the Great Famine, but by the mid-1850s it was reduced to limited numbers along the north coast and by 1900 it had become so rare that it was described as only a casual visitor to Ireland.[91] The last pair bred in County Derry in 1886 but shooting by gamekeepers and farmers as well as the widespread use of poison, laid to kill predators, led to its demise. In the 1950s, the buzzard population in Ireland had begun to recover with a handful of nesting pairs, but only one nest was confirmed as successful. Then myxomatosis, a viral disease infecting rabbits, was intentionally introduced into Britain and Ireland. The rabbit is one of the main prey items of the buzzard and so the species became extinct from Ireland for the second time in the late 1950s due to the decline in rabbit abundance.[92] I had never seen a buzzard before I worked in Northern Ireland in the 1970s, but by then, a few birds were already spreading back into the north coast area, probably from the Scottish population. By 2007–11 they were widespread in the east of the country with a reduction in persecution and poisoning. [93] Almost every day now I see at least one buzzard soaring over our farm in County Wicklow or emerging from local patches of woodland. They are now found mainly where the landscape is a mixture of small woods and hedgerows for nesting and farmland where they feed. The buzzard is unique among the birds of prey in that it became extinct twice in Ireland, but managed to return and recover by itself, and it is now common and widespread.

Confined to coastal areas, the gannet is the largest seabird breeding in Ireland. In past centuries it was extensively hunted as the chicks are large and remain on their cliff nesting sites, where they are very vulnerable, until August each summer. In the 19th century, fishermen from the Blasket Islands and Dingle, County Kerry, would sail as far as the Skellig Rocks to slaughter the gannets. Islander Tomás Ó Crohán related how 'They sprang up it and fell to gathering the birds into the boat at full speed and it was easy to collect a load of them, for every single one of those birds was as heavy as a fat goose.'[94] The gannets were dried and eaten in the winter when food was scarce, while the feathers were sold to pay the rent. Since this subsistence hunting died out in the early 20th century, the gannet population has recovered steadily from about 1950 onwards. I have great memories of circling the largest Irish colony at the Little Skellig in County Kerry with gannets diving into the sea all around our boat.

There is a long tradition of counting gannets and the changes in numbers are better documented than for any other seabird in Britain and Ireland. Many of the smaller colonies are counted each year and censuses of the whole population have been made at intervals since the early 1900s when a total of about 20,000 nests was estimated.[95] By 2014, the gannet population had reached nearly 48,000 occupied nests in Ireland with a total of six colonies, three of which were newly established since the 1980s.[96] This increase has happened right across the range of the gannet in the north-east Atlantic and it is considered to be a recovery from prolonged and extensive hunting. However, gannets are not completely out of trouble yet. They depend totally on shoaling fish such as herring and mackerel to survive and the increasing levels of overfishing directly threaten their livelihood.[97] More recently, the gannet is one of the wild bird species most severely hit by bird flu, although the full impact of this on the population may not be known for some time as young birds take several years to mature and return to the colonies.

The cessation of hunting also benefitted the grey seal in these same waters. The white-coated pups are confined to land for the first few weeks of life and so they were easily clubbed or shot by coastal dwellers in previous centuries. The oil-rich meat was often dried for the winter while the skins made excellent waterproof coats for fishermen and were occasionally used for covering the hulls of small boats. In Britain, large-scale culls of grey seals in the North Sea, Orkney and the Hebrides were carried out in the 1960s and 1970s as population control measures. Subsequent monitoring shows that since legal protection was given to seals in 1970, numbers have increased consistently. In Ireland, up to the 1970s, a bounty was paid for the head of every seal turned in by a fisherman as it was considered that they were limiting the catch of valuable fish like salmon. However, new legislation put an end to this and the grey seal population began to expand, especially breeding on now-deserted islands such as Great Blasket and the Inishkea Islands. In 2017–18, almost 4,000 grey seals were counted in Ireland with the total population now nearly three times higher than the 2003 count.[98] The total number of grey seals in the UK in 2019 was estimated at over 157,000 animals, far eclipsing the Irish population.[99] As one of the largest mammals in this country, it is now returning to its former haunts as a result of the reduction in disturbance and culling.

I can hear the calls of the great spotted woodpecker without leaving my front door and their distinctive drumming on dead branches is a sure sign of breeding. But just a few years ago this was unimaginable in Ireland. The only evidence that they had ever bred here came from two small bones of the great spotted woodpecker that were found in caves over a century ago in two locations in County Clare, suggesting that they were present in the Bronze Age.[100] Because they are so common and widespread in the neighbouring island of Britain, it had always been assumed that they had died out with the wholesale clearance of our native woodlands over the millennia. Then, in a dramatic discovery, a juvenile bird was seen in a garden in County Down in 2007, but it took a further two years for the first occupied nests to be found in woodlands in County Wicklow. Genetic evidence has shown that they originated from the British population, which had been expanding at that time.[101] After that, the expansion in their Irish range was rapid. By 2023, the Irish Biodiversity Data Centre reported that they had spread widely across the east and midland counties and will probably colonise suitable woodlands throughout the country.

New species of insects sometimes appear spontaneously in Ireland, having never been recorded here previously. The ivy bee has been spreading rapidly across Europe. Unlike bumblebees that nest in colonies, this solitary species makes single nests in light or sandy soil on southern-facing banks and cliffs with ivy nearby for foraging. It feeds mainly on the nectar of ivy flowers and can be seen from early September to early November when this plant is in bloom. It was first recorded in England in 2001 and made an appearance in Ireland in 2021, with the discovery of a colony of approximately a thousand ground nests at the Raven Nature Reserve, County Wexford. With help from volunteers, Professor Francis Ratnieks surveyed the distribution and abundance of the ivy bee, recording it at 91 locations in the coastal areas of Wexford and Wicklow.[102] Other insect species are now colonising Ireland due to warming temperatures and more are likely to arrive, a trend that is almost certainly driven by climate change. Some other conspicuous examples include the comma butterfly and golden-ringed and emperor dragonflies. Typically, new insect species are first found on the east coast of Ireland, when there are sufficiently large populations in Britain or Continental Europe and when the weather conditions are favourable for dispersal.

There are some native plants that also seem to be returning of their own accord. Sea kale is a deep-rooted perennial plant of maritime shingle which is occasionally cultivated as a vegetable and may well have been eaten by past coastal communities. It became quite rare in previous centuries, but since 2000, it has been recorded in many new sites, and some former ones, with recent records from around much of the Irish coast. It is unlikely that such a large and obvious plant was overlooked by earlier botanists. Either a warming of the climate may have given rise to a temporary spike in numbers, or there may have been a real permanent increase. However, this species is still listed as near threatened in the Ireland Red List and is a priority species for conservation in Northern Ireland. A slender annual plant, thale cress, was once a scarce native of rocky places with thin soils, but over the last 50 years it has spread massively and is now an extremely common weed of gardens, wall bases, gateways and other ruderal habitats. It seems to thrive in sites where herbicide is applied regularly, but this does not prevent the plant from germinating and setting seed between treatments. It is not known whether this increasingly common plant came from native or introduced sources.[103]

The wild plants and animals that have become extinct or extremely rare in Ireland are mainly those, such as the freshwater pearl mussel, with quite specialised habitat requirements that are no longer available. Others, like the eagles, suffered from sustained hunting or persecution which they could not survive. Those that survived are species, such as the badger, which have more general needs and can adapt to new habitats, including urban environments. Encouragingly, when the pressures are removed or their habitats are restored, some of these animals, like the buzzard, can recover by themselves, while others are colonising Ireland from further south due to climate warming. These are the winners. Other lost species need active help to reestablish them here. For a few, such as the great auk, which is lost from the entire planet, extinction is for ever. These are the real losers.

Bringing them back

Almost every day, I watch the beautiful, effortless flight of a red kite circling over our farm in County Wicklow. Occasionally it drops to the ground to investigate a potential source of food or to hunt for earthworms in a ploughed field. A few years ago, this would have been unimaginable, as these birds disappeared completely from Ireland about three hundred years ago. Their reputation for taking poultry and young game birds did not do them any favours and they were persecuted by shooting and poisoning until they became extinct.[104] The first reintroduction of these beautiful birds began in 2007. Wicklow was chosen for the release site as it closely resembles the wooded valleys and small farms of central Wales where the last native red kites in these islands survived. I was watching with excitement when the cage doors were opened and the young birds took to the skies of Ireland for the first time in several centuries. They adopted their new home immediately and now there are well over a hundred pairs of kites breeding in the east of Ireland in one of the most successful nature restoration projects ever undertaken in this country.

Many plant and animal species are unlikely ever to return to Ireland or recover from past losses without some help. This may be because they have poor dispersal abilities, cannot cross the sea from Britain or the European continent or may be so specialised that they require habitat conditions that are now rare or completely absent in Ireland. But there have been a few successes in recent years. Before the reintroduction of a native species is approved, it has to meet a number of strict criteria. These usually follow the guidelines published by the International Union for Conservation of Nature. In their definition, 'Reintroduction is the intentional movement and release of an organism inside its indigenous range from which it has disappeared.' This must be intended to yield a measurable conservation benefit at the levels of a population, species or ecosystem and not only provide benefit to individual animals or plants.[105] Normally a detailed feasibility study is undertaken first along with an assessment of the risks to other species and to local human populations.

Birds of prey are a group of species that have suffered greatly in the past through illegal poisoning, persecution and egg-collecting. Many of the larger species such as the eagles disappeared completely from Ireland in the early 20th century. Founded in 1999, the Golden Eagle Trust is a charity dedicated to the conservation and restoration of Ireland's native birds and their habitats, in particular declining, threatened and extinct species. So far it has managed reintroduction programmes for golden eagles in County Donegal, white-tailed eagles in County Kerry and red kites in County Wicklow, in partnership with the NPWS. The project to bring back white-tailed eagles got underway in 2007. These are simply magnificent birds with a massive yellow hooked bill and huge wings spanning well over two metres. Over several years, a hundred young eagles were collected under licence from nests in Norway and released in the Killarney National Park in Kerry. After release, some of the young birds hung around Kerry but, within a few months, they were travelling further afield exploring their new country and frequently moving from county to county. The first nesting pairs were recorded in 2012 and the first Irish-born white-tailed eagle chicks fledged successfully in 2013.

I went to Mountshannon, on the western shores of Lough Derg, to search for one pair of wild eagles and, with the aid of a telescope, I was able to watch these impressive birds flying out across the lake from their nest on a wooded island. The rich fishing here has clearly been a key factor in the success of the birds, and the local community has welcomed and supported the project. By 2018, with at least 12 pairs holding territories in Ireland, the eagles were breeding across four counties (Kerry, Cork, Clare and Galway) and were showing signs of expanding their range again, into the north-west and midlands.[106] In 2020, a new phase of the reintroduction programme began with the release of young eagles at several sites, including Lough Derg, the Lower Shannon Estuary and the Lakes of Killarney. One of these eagles was recently shot in County Roscommon, so negative attitudes to harmless predators still remain in some places. Up to 42 Irish-bred young eagles had fledged from wild nests by 2022 but productivity and the number of fledglings were still low compared with more established populations elsewhere. Restoring a top avian predator and scavenger requires long-term commitment to be successful.[107]

Some extremely rare species have been monitored very carefully over many decades and, in some cases, centuries. One of these is a beautiful plant called cottonweed, whose only known site in Ireland and Britain is a shingle barrier beach on the south Wexford coast. I have seen it there, its grey and yellow flowers shining out among the surrounding marram grass. In the early 20th century, various botanists found that it was relatively secure in several locations, but a century later it had drastically declined with only a handful of plants remaining at Lady's Island Lake. The main reason for the decline in the population is thought to be competition with marram grass which has spread throughout the area and effectively outcompeted it. The underlying reasons for the increase in marram are not fully understood but likely involve a combination of factors, including changes in the type of sediment deposited and in the frequency of sand deposition by storms. Various efforts were made to rescue the cottonweed by clearing

marram around the survivors and by collecting seed of
cottonweed and planting it in other nearby sites such
as the barrier at Tacumshin Lake. This achieved mixed
success, so in 2013 the NPWS decided to collect further
seeds and cuttings which were grown to seedling stage in
controlled conditions at the National Botanic Gardens
in Dublin. Several years later, these were planted out in
a specially cleared and fenced area at Lady's Island Lake
where coarse sand and gravel had been added to the
surface. So far, the results have been very encouraging
with a 60 per cent success rate after four years. This
could almost be described as restoration gardening. It
shows that, with a concentrated effort underpinned by
good scientific data, a seriously threatened species can
be rescued before it is too late.[108]

More often, rescue operations for threatened plants are
initiated because of some direct threat from development.
Translocation is occasionally used to move a rare species
into an alternative place out of the way of development
or to ensure that all the eggs are not in one basket. The
rare and legally protected triangular club-rush is recorded
in Ireland only in the estuary of the River Shannon and a
few of its tributaries (the Maigue and Owenagarney rivers)
in counties Limerick and Clare. There has been a dramatic
decline of the species in Britain. The Irish naturalist Robert
Lloyd Praeger mentioned the species in his famous book *The
Way That I Went*. 'Between the bridges, it forms tall dark
groves, much resembling those of the common bullrush.'[109]
With a friend, I canoed along muddy creeks between tall

reeds in the estuary to find and map this rare plant. Here we found plenty of the club-rush growing right down at the tide edge, but when we made it back to shore, the tide had gone out, leaving 50 metres of soft, gloopy mud to navigate as we dragged the canoe ashore. The development of the Limerick Southern Ring Road, including the Limerick Tunnel, would have impacted populations of this rare plant within the Shannon Estuary so some of the threatened plants were removed from the line of the proposed road and stored in specially constructed holding tanks. Two years later, following construction of the tunnel, plants were relocated to the estuary with the aim of re-establishing the species in the locations from where they had been moved. The relocated plants were monitored annually between 2009 and 2014 and it was clear that they had responded by expanding significantly.

Many species of freshwater fish were introduced to Ireland since the Normans arrived, but reintroductions were less common. The only truly native fish are those which spend part of their life cycle in freshwater and part in the sea, like salmon, trout, eel and lamprey. In the excavations at the Mesolithic settlement at Mount Sandel, County Derry (from around 9,600 years ago), four-fifths of the animal bones found were from fish, mainly salmon, trout and eel. The 12th-century Welsh writer Giraldus Cambrensis noted that, 'The rivers and the lakes are rich in fish peculiar to them and especially in fish of three kinds, namely salmon, trout and mud-eels. The Shannon abounds in sea lampreys. They serve as luxuries for the rich.'[110] Most coarse fish are believed to have been introduced between the 12th and 15th centuries. Pike was first recorded in the late 13th and early 14th centuries. Other introductions included gudgeon, rudd, tench and stone loach which are all considered benign additions to the Irish freshwater environment. Less welcome are invasive species such as bream, perch and roach which are likely to displace native species. More recently, there were introductions in the 19th and 20th centuries of rainbow trout (generally unsuccessful) and dace (to the Munster Blackwater).[111] Although the evidence is scanty, Dr Julian Reynolds believes that the white-clawed crayfish was introduced by early people, most probably by the Cistercians some decades before the Norman invasion. This is supported by genetic evidence. Reynolds says, 'I've translocated them from Meath and Westmeath to former crayfish lakes including White Lake and Lough Lene. These reintroductions were initially successful.'

Captive breeding is sometimes used as a last resort to save species that can be released back into the wild when habitat conditions improve. Internationally, there have been a few high-profile cases with mammals and birds such as the Californian condor, but success is rare. The Nore pearl mussel occurs only in one river system in the south-east of Ireland and it is classified as critically endangered, mainly due to a drop in water levels and serious declines in water quality. To replenish the stocks there was some captive breeding of the mollusc between 2005 and 2014. To date, and in spite of huge efforts, there are no successful results for captive breeding leading to recovery or survival of mussels in the wild, and nor is there any evidence of long-term success for other freshwater pearl mussels in Ireland. Of course, captive breeding is expensive and does not restore the habitats of the animals. Dr Evelyn Moorkens, an international authority on the pearl mussel, says,

> I think it is true for most captive breeding pro-grammes, not just mussels, that this approach is only useful in preventing the extinction of genetic units to buy time until real [habitat] restoration is undertaken. The problem with European proj-ects to date is that the restoration has not been achieved at the same speed as the captive breeding. A key problem in keeping captive-bred freshwater mussels is poor genetic diversity. When river con-ditions are bad, juvenile mussels need to be reared up to 11–12 years of age before they can be placed in the river but very few juveniles survive to this age in captivity. It has been found that in their early stages in captivity, genetic diversity is high, but the surviving older juveniles have only a frac-tion of that diversity.

It is clear that the future will not be secure for the freshwater pearl mussel without a much greater effort in habitat restoration, particularly water quality improvement.

Also living in freshwater, the natterjack toad is one of only three amphibian species in Ireland. Unlike frogs, they do not hop but rather walk along on all fours. They spend their adult lives in coastal, sandy areas, while breeding in warm, shallow ponds. In Ireland, the natural range is limited to a small area on the Dingle Peninsula in County Kerry where they are found in some ponds in sand dunes and others in farmland. There is a small population in south-east Wexford, relocated from Kerry in the 1990s. The species' range is estimated to have contracted by over 50 per cent between 1900 and the 1970s.[112] As a result, the toad is classified as being 'endangered' in Ireland.[113] The decline in farmland ponds due to wetland drainage and agricultural intensification is the main cause of natterjack toad decline. Survival rates of eggs and tadpoles in the wild are typically very low due to predation and drying out of the ponds in summer.

To help boost the natterjack population, a 'headstarter' programme was initiated by the NPWS. This is a kind of captive breeding project which involves the transfer of toad spawn, laid naturally in pools in County Kerry that were drying out, to both Fota Wildlife Park in Cork and Dingle Aquarium in Kerry. The spawn is incubated in tanks before being released back into the native area in Kerry as toadlets. Since 2016, the two aquaria between them have released about 10,000 toadlets back into the wild. This captive rearing has proved very successful and a modest programme of translocations has also taken place helping to supplement a very vulnerable population of this species. All of this depends, of course, on suitable habitats being available in the wild. To ensure this, the

NPWS has offered grants to local farmers to dig more than a hundred ponds, all within the former range of the species, and to maintain them in a suitable condition for toads through hand clearance of vegetation and grazing the surrounding sward.

At one level, species reintroductions or translocations excite romantic feelings in many people and create a feel-good factor in our attempts to restore animals and plants lost from the distant past. But these methods are often used as a last resort when all other restoration attempts have failed. On another level, they can be vital in restoring critical components and functions in impoverished ecosystems. With sufficient scientific research, financial backing and buy-in by local communities, reintroduction or rescue of threatened or extinct native species can be successful and can contribute significantly to restoring balance in the landscape. The choice of which species should receive the limited resources available is based on a number of scientific and logistical criteria. There are some plants and animals that are regarded as keystone species and should get greater priority. As in the building of a stone bridge, these are crucial to support many other species. If they are removed, like the keystone in a bridge, the whole ecosystem may collapse or become so simplified that it is unstable. This illustrates how interconnected all the species are in a fully functioning ecosystem.

Probably the most celebrated example of a keystone species is the wolf in Yellowstone National Park in the USA. Established in 1872, this was the first national park in the world, but by the mid-1900s wolves were absent here, having been 'cleaned out' across many areas of North America where they were seen as a threat to farming and humans generally. Some ecologists began to notice that this absence was having detrimental impacts throughout the National Park. Elk had increased and their browsing had mainly caused the decline in river cottonwoods and upland aspen woodlands. Increased populations of intermediate predators such as coyote caused declines in other smaller animals. With greater awareness of the impacts and more public support, an initial group of 66 wolves was released in the Park in 1995 and ten years later the wolf packs had between them reached well over 300 animals. The beneficial effects on the park ecosystem were clear to see. Elk numbers declined but, more significantly, their grazing patterns changed as they avoided dense tree cover to lower the risk of ambush by wolves. This led to the recovery of aspen, willow and cottonwood trees. Released from grazing, the enhanced growth of the riverside woodlands led to the recovery of aquatic life, and the increase in wolf kills benefitted a whole range of scavengers such as grizzly bear, cougar, wolverine and raven. This natural process is known as a 'trophic cascade', where one keystone species can bring about a recovery in multiple habitats and species. This example played a key role in the emergence of restoration as a new conservation tool. With a bit of assistance, it showed that nature can recover much of its biodiversity quite quickly.[114]

In Ireland, the pine marten may turn out to have a similar role in the woodland ecosystem. These are omnivores that feed on a wide variety of different foods: birds' eggs and nestlings, fruits in autumn and small mammals. As expert tree climbers (the Irish name is *cat crainn*, the tree cat), they are well able to hunt squirrels either in the treetops or on the ground. The decline and near extinction of the pine marten in Ireland due to persecution coincided with the introduction and spread of the introduced American grey squirrel which ultimately led to the decline of the native red squirrel. However, better control of poisoning and persecution combined with the spread of forestry in the late 20th century led to a marked expansion of the pine marten and a corresponding decline or even complete disappearance of the grey squirrel in some parts of the country. It seems that the grey squirrel is especially vulnerable to this predator because it did not evolve with it as the red squirrel did and because the greys spend more time feeding on the ground. It is also known that grey squirrels will leave an area if they get even the scent of a pine marten. It has also been found that the red squirrels have returned in areas where the pine marten is now present. However, grey squirrels are predicted to survive in refuges such as parklands in and around towns and cities, which the pine marten avoids.[115]

Scientists are beginning to debate whether other keystone species could have similar beneficial effects for biodiversity in Ireland. European lynx are known to have occurred in Ireland in the Late Stone Age (Neolithic period) although they may have been introduced here by early people. This small wild cat is about the size of an Alsatian dog and can have similar benefits to wolves for ecosystems in many countries. The lynx is a solitary, shy hunter requiring large areas of forest cover from which to launch ambush attacks on smaller deer, its favoured prey. Across mainland Europe, the Eurasian lynx is staging a comeback. Freed from the pressures of unsustainable hunting and benefitting from a softening of public attitudes, this wild predator is now increasing in numbers and expanding its range. A series of lynx reintroductions have taken place since the early 1970s in Switzerland, France, Germany, Italy, Austria, Slovenia, Poland and the Czech Republic. Not all of these projects have been successful, but useful lessons have been learnt from the failures as well as the successes.

There have been several studies to evaluate the ecological feasibility of returning lynx to Scotland and, with expanding forest habitat and abundant prey, there is growing support for their return there. These are essentially forest animals, but they may occasionally prey on livestock such as lambs and gamebirds such as grouse which has led to significant resistance from farmers and other landowners. In 2021, the Lynx to Scotland partnership carried out a comprehensive study to accurately evaluate people's attitudes towards bringing back lynx.[116] The study revealed wide-ranging views from stakeholders, including farmers, gamekeepers, foresters, conservationists, landowners, tourism operators and rural communities. It showed that views about lynx reintroduction are far more diverse, nuanced and complex than a simple 'for' and 'against'. For example, browsing by deer in forests is a major problem for the reestablishment of native woodland in both Scotland and Ireland. The potential for controlling or reducing the impact of invasive species like sika deer is significant. However, as wide-ranging predators, each lynx requires a very large hunting territory and they naturally occur at low density. Males can have home ranges of over 200 square kilometres, far bigger than those of our existing carnivores such as otters, foxes and pine martens. Wild forest areas of this size are simply not available in Ireland. It could be many years before Ireland is ready for even a trial reintroduction of lynx.

The first appearance of human settlers in Ireland coincided with the introduction of a number of other woodland mammals which would have been useful to the early people as food, skins and bones. This time is termed the Mesolithic period (or Middle Stone Age) and it lasted until about 6,000 years ago. The larger animals, such as wild boar and badger, were probably brought in intentionally as were lynx and wild cat which would have been valued for their fur, for clothing and for other domestic uses. One can imagine these animals being transported across the sea in primitive dugout canoes to Ireland where they were released or escaped into the still-extensive forests that covered the country at this time. Smaller woodland mammals such as wood mouse and pygmy shrew were probably transported unwittingly, travelling in the bottom of boats among bedding or food stores. The first farmers arrived in the Neolithic and they brought with them red deer, presumably as quarry to hunt, as well as domestic animals such as primitive cattle.

The wild boar is a widespread native mammal in Europe and there is plenty of archaeological evidence that it occurred in Ireland from the Mesolithic right up to medieval times. It is very likely that they were introduced here for hunting by the earlier settlers and they would also have been an important prey species for wolves. However, the clearance of the ancient forest cover and hunting pressure led to their extinction. In other parts of Europe, boars eat mainly plant material, especially roots and tubers which they dig up from the woodland soil and tree seeds such as acorns fallen from oak trees in the autumn. They also consume green buds and shoots of young trees. This disturbance and browsing could be both beneficial and detrimental to the woodland ecosystem but the species certainly had a key role in supporting other species which hunted them. The wild boar's rooting behaviour helps nutrient cycling in a woodland and can create fertile conditions for tree seed germination – particularly Scots pine and birch. By digging into the soil, they expose invertebrates for many bird species, while creating patches of bare soil for burrowing insects such as mining bees and some species of beetle. Their ploughing actions can be a boon for biodiversity, facilitating the colonisation of more plant species. Their digging can help to keep dominating shrubs such as bracken under control, giving other plant species a greater chance of growth, and their need for mud baths in the summer can create miniature ponds that attract amphibians and dragonflies. However, too many boars could cause significant problems for agriculture and they can be quite dangerous if people get too close to them in the confined spaces of woodlands.

Since 2009, some boars have been unofficially released in Ireland although most of these are regarded by scientists as hybrids with domestic breeds and are thus described as feral pigs. These animals are classed as invasive and could do damage to agricultural crops and transmit various diseases to livestock.[117] Most were shot on sight and genetic analysis of a sample of animals showed that the majority were related to domestic breeds while only three were classed as wild boar hybrids.[118] The wild boar remains something of a fugitive in the countryside today. It is not recognised as a wild native species so cannot be reintroduced officially.

The return of the osprey is a well-known success story in Britain. Once exterminated, they returned of their own accord to Scotland and have since spread widely with help from conservation bodies. By moving young birds to other areas, they have become established on a number of lakes and reservoirs throughout Britain where there are now over 240 breeding pairs. There are similar reintroduction programmes in many other European countries.[119] The osprey was once common in Ireland. This formidable bird of prey, which has a wingspan of up to 180 centimetres, feeds exclusively on fish and hunts in both freshwater and saltwater. Bones of this species, dating from the 10th–11th century, were found in excavations at Fishamble Street, Dublin City, suggesting that they may have bred near the River Liffey. Giraldus Cambrensis gave a very accurate description of the bird and its hunting behaviour in his *Natural History and Topography of Ireland* in the 12th century. There are many other later mentions of the osprey but by the late 18th century it was already in steep decline and a century later it was recorded only as a rare visitor.[120]

The Scottish and Scandinavian breeding ospreys migrate through Ireland to their wintering grounds in West Africa. I have vivid memories of watching one hunt for mullet on the small estuary of Broadlough in County Wicklow. In 2022 the NPWS announced that it was working with its counterparts in Norway to prepare for reintroduction of the osprey here after an absence of about 250 years. The following year, the first nine osprey chicks in this programme were flown in from Norway and released into the skies in County Waterford. Over the next five years, between 50 and 70 osprey chicks will be reintroduced. The hope is that they will establish a breeding population here and will then return every year after their long migration from Africa in the spring. Given the successful reintroduction of white-tailed eagles to this country and their similar preference for hunting in lakes, the prospects for the osprey are good. The suitability of Irish lakes was finally confirmed when the dramatic news emerged that a single pair of wild ospreys had bred in 2023 on Lough Erne, in County Fermanagh. The charity Ulster Wildlife, which had been observing the same birds for three full seasons, said that 'These distinctive birds of prey had recolonised naturally in the area and successfully produced at least two, possibly three chicks – the first known wild osprey chicks on the island of Ireland in modern times.'[121]

While there is currently no archaeological evidence that European beavers previously occurred in Ireland, their reintroduction in Britain has had marked benefits for aquatic ecosystems. On their first reintroduction to Scotland, they immediately began to fell trees and create dams, thus raising the water levels to flood large areas, creating ponds and wetlands while restoring floodplains. Studies in southern Sweden have shown that, in areas where beavers are found, the richness of plant species is up to one-third higher than in wetlands unoccupied by these animals. The effect of these ecosystem engineers is to increase the complexity or 'patchiness' of the habitat, radically transforming uniform rivers and lakes and the species that they support. A UK study found that the abundance of freshwater invertebrates – dragonflies, damselflies, diving beetles, water boatmen and backswimmers – was three times higher in beaver ponds than in unmodified agricultural streams. In reshaping rivers and wetlands, their dams have been shown to improve downstream water quality and reduce peaks of flooding and drought. [122] Could beavers do the same for our damaged rivers and streams here?

Some lost species can expand again of their own accord with improving conditions such as restored habitats or a reduction in the pressures that caused their loss in the first place. But others require some focussed assistance to return if they are unable to cross the seas to Ireland. The benefits of restoring lost species to damaged ecosystems are plain to see in other countries and this may form a key part of the restoration effort here in future. However, great care needs to be taken to ensure that conditions are now sufficiently improved and that there is enough habitat available to support these species. Potentially negative impacts on local communities and on enterprises such as farming and forestry also have to be carefully considered. However, the return of an extinct species such as the red kite to its former haunts is usually widely welcomed as a much-needed success story in an age of widespread loss of biodiversity. The English journalist George Monbiot wrote, 'Ecologically they could live here today. The obstacles are cultural and economic.'[123]

Restoring the future – a noble goal

This book addresses the question of whether there is a future for nature in Ireland, given the severe declines of many habitats and species in recent times. The old policies and practice of nature conservation have largely failed to stem the tide of loss. It is time to actively restore at least some of the damaged habitats and lost species and to set recovery in a more sustainable direction. International political thought is changing and the question is whether Ireland can respond with commitment and action. We know what needs to be done. We know how to do it. There are already templates in this country and elsewhere for how to restore nature while living alongside it. This is the time to secure a future for wild nature.

In 2003, the late Michael Viney wrote in the preface to his scholarly book, *Ireland: A Smithsonian Natural History*, that

the next decade or two could be a difficult time for the Irish landscape and its wildlife, as pressures flow out from a totally unprecedented economic boom, a rising population, and even more tourists seeking the 'unspoiled' corners of the island. The impetus for conservation, substantially nourished by Ireland's membership of the European Union, has yet to find a full endorsement in the Republic's political soul.[124]

He was right, of course. But to that list of substantial pressures he could have added the growing impacts from climate disruption to life on land and sea. Now, two decades later, we can say that the 'difficult time' for the Irish landscape has become more than a temporary situation. In 2019, a National Biodiversity Emergency was declared by the State, but this has made little difference to the urgency of the response.

Over the last half-century, I have seen interest in the environment grow from a fringe or even eccentric activity to something more central in everyday life. One example of this is the uptake by farmers and urban dwellers of action for insect pollinators. Interest in nature and wildlife is growing in Ireland and this may be partly a result of the greater proportion of the population living in urban areas who have lost a direct connection with the natural environment. To translate this into political action and ultimately finance to ensure the recovery of nature from the pressures of modern land use will require much more action from citizens.

In the spring of 2023, the Citizens' Assembly on Biodiversity Loss published its much-anticipated report. The result of a full year of deliberations by a hundred randomly selected citizens, it tells us the bad news first.

Only 2 per cent of the country has native
woodland. Over a quarter of Ireland's regularly
occurring bird species are in danger of extinction.
At least one-third of protected species are
declining in population. Almost 30 per cent
of our semi-natural grasslands have been lost
in the last decade. Less than half of our marine
environment can be described as healthy. Over
70 per cent of our peatlands are in bad status
and only a small fragment remains intact. The
majority of our agricultural soil is in a suboptimal
state, contaminated by nitrates and phosphates.
Most worryingly, our water quality – the very
foundation of life – is continuing to decline, with
almost 50 per cent of freshwater systems in Ireland
in poor and deteriorating condition.[125]

After deliberating on how the state can improve its
response to the issue of biodiversity loss, the Assembly
agreed 159 recommendations. It called on the Oireachtas
to accept these recommendations and implement them
without delay in order to curb the crisis of biodiversity
loss and allow for the conservation and restoration of
biodiversity for the people of Ireland, present and future.
The recommendations are both wide-ranging and incisive.

It would be easy, however, to be cynical. I have seen
many such government-appointed groups come and
go, from the Wildlife Advisory Council of the 1980s
to the National Biodiversity Forum of the 2020s,

with many good and expert people making worthy recommendations that were simply put on a shelf to gather dust. Ultimately, it is a question not of fine words, targets and aspirations, but of action on the ground and putting a brake on some of the mad dash for economic 'growth' that seems to grip successive governments in this country. The chairperson of the Citizens' Assembly, Dr Aoibhinn Ní Shúilleabháin, summed it up neatly in her foreword to the report.

> It is time we start valuing our natural heritage as much as our cultural heritage, start treating our bogs like our Book of Kells, value our rivers and coastal waters as much as our multinationals, and cherish our forests as a part of our living history.

But there is much more required than simply putting a value on nature. The functioning of the natural environment in this country has been so badly damaged that it will take active restoration work to reach a place where nature can take care of itself. Current progress in restoring nature is uneven across the habitats. Rehabilitation of cutaway lowland bogs is well underway and work has started to rewet some damaged mountain peatlands. There is increased focus on the planting of new native woodlands, but the scale is tiny. Many of the surviving fragments of old woodland continue to deteriorate due to lack of appropriate management. While there are some isolated examples of a more enlightened approach to 'farming with nature', intensive agriculture and forestry continue to impact our rivers which are suffering alarmingly. Overfishing remains the most pervasive threat in the sea and the establishment of meaningful marine protected areas has been painfully slow. The reintroduction of some lost species such as the larger birds of prey has shown what can be achieved with long-term commitment, but few other extinct species have received much attention. Valuable though they are, these restoration projects are small in scale. Meanwhile, the rest of the landscape and the wildlife that it supports are becoming more impoverished year by year.

The Society for Ecological Restoration is an international science-based body that debates and refines the practice of nature recovery. It has published quite detailed principles to provide consistency for projects across the world. According to the Society, ecological restoration:

- engages stakeholders,
- is informed by native reference ecosystems,
- draws on many types of knowledge,
- supports ecosystem recovery processes,
- is assessed against clear goals and objectives, using measurable indicators,
- seeks the highest level of ecosystem recovery possible,
- gains cumulative value when applied at large scales,
- and is part of a continuum of restorative activities.

At a more practical level, there are certain guidelines that should be followed when undertaking a restoration project. These were outlined by the visionary conservationist and founder of the UK organisation Trees for Life, Alan Watson Featherstone.[126] He said that we need to work from areas of strength – the areas where the ecosystem is closest to its natural condition. One example of this would be the practice of reafforesting the uplands by protecting from grazing the small groups of trees in gullies and river corridors, thus creating a source of tree seed to spread out to the wider slopes. His advice was to pay particular attention to keystone species – those on which many others depend. These can be iconic predators like the pine marten or much less conspicuous species like the *Sphagnum* mosses which are essential for the regrowth of bogs. Some species with limited ability to disperse, such as the aspen tree, freshwater pearl mussel or orchids, need very particular conditions to be restored. We should only reintroduce native species that are unlikely or unable to return by themselves.

Watson recommends a focus on reestablishment of ecological processes such as natural succession and the use of pioneer species to facilitate the restoration process, mimicking nature wherever possible. We should recreate ecological niches where they have been lost, provided that we know where they existed. For example, the retention of dead and decaying wood in a forest provides niches for hundreds or even thousands of plants and animals. He talks about re-establishing ecological linkages

– reconnecting the threads in the web of life. Many flowering plants depend on pollinating insects to produce seed and without them a semi-natural grassland will be significantly impoverished. Re-establishing essential ecological processes, such as predator–prey dynamics, is another of Watson's principles. The natural processes of erosion and accretion in sand dunes can sometimes be restored simply by removing the artificial barriers, or 'armouring', which separate them from beaches. The control or removal of introduced invasive species is a difficult and often intractable problem but, where it can be achieved, there will be significant benefits. Fencing out non-native deer, or reducing their numbers, can result in a spectacular recovery of natural vegetation and all the animals which depend on it. We can also remove or mitigate the limiting factors which prevent restoration from taking place naturally. These factors can be as diverse as poisoning with chemicals, overgrazing and burning of the hills, drainage of rivers or overfishing in the seas. Finally, Watson says that nature will do most of the work if we give it half a chance.

The Irish author Paddy Woodworth, in his ground-breaking book *Our Once and Future Planet*, asks the question: 'Why restore?' He answers it by saying that, 'There is something badly off-kilter in our relationship with the rest of the natural world and ecological restoration offers unique and refreshing perspectives for setting that relationship on a better course.' He rightly points out that, 'The environmental movement has generally tended to focus on catastrophic scenarios, on narratives of loss and desolation.'[127] By contrast, I have seen how nature restoration has the potential to inspire people, to give them hope and to reset our course to a more positive form of action. I have seen the excitement that urban-dwelling people get from planting a few native trees or digging a pond. It can reconnect them with nature and confirm how a little positive action will bring us back to a better relationship with it. On a larger scale, it is hard to ignore the interest generated across the world by the reintroduction of wolves to Yellowstone National Park in the USA or the conversion of Knepp Estate in England from intensive farmland to semi-natural habitats which quickly became colonised by native species.[128]

Half a century after Ireland joined the EU, gaining significant benefits as a member state, the European Union has agreed a new law to restore ecosystems for people, the climate and the planet and this has now passed into law. This is the first continent-wide comprehensive law of its kind. It is a key element of the EU Biodiversity Strategy, which calls for binding targets to restore degraded ecosystems, in particular those with the most potential to capture and store carbon and to prevent and reduce the impact of natural disasters. Europe's nature is in alarming decline, with more than 80 per cent of key habitats in poor condition. Restoring wetlands, rivers, forests, grasslands, marine ecosystems and the species they host will help to increase biodiversity, secure the things nature does for free, like cleaning our water and air, pollinating crops and protecting us from floods, and limit global warming using nature-based solutions, while building up Europe's resilience by preventing natural disasters and reducing risks to food security. The new law combines an overarching restoration objective for the long-term recovery of nature in the EU's land and sea areas with binding restoration targets for specific habitats and species. These measures should cover at least 20 per cent of the EU's land and sea areas by 2030, and ultimately all ecosystems in need of restoration by 2050. There is no doubt that collaboration with rural dwellers, farmers, foresters and those in the fishing community will be vital if nature restoration is to succeed. There will be a requirement for large numbers of skilled machinery drivers, fencing specialists, tree-planting workers and ecologists to form the workforce for the massive restoration task ahead. Training and adequate payment for this work will help to replace the income that these people have previously had from conventional land use. I have no doubt that, with some well-publicised examples of how nature restoration is good for the climate, the land and all its inhabitants and users, public support will increase. Like the planting of a tree, the best time to do this was 20 years ago. The restoration work needs to start immediately.

Speaking about the cause of protecting nature in 1948, one of Ireland's most eminent naturalists, Robert Lloyd Praeger, said on the radio,

> We are at the beginning of a long and also delicate piece of work, calling for patience, tact, judgment and industry, as well as enthusiasm; but our goal

is a noble one, and once it is fully appreciated there is little reason that anyone's hand should be turned against us.

Nature restoration is now the way forward. Praeger's 'noble goal' is achievable but, as in sport, it will take a team effort. Despite decades of often disappointing efforts to try and stem the losses of nature in Ireland, I remain optimistic. The new generation of young people are better informed about environmental issues than I was at their age. They are more confident and generally have good communication skills. They are stronger and healthier. And some of them are more committed to taking action for nature than any of their parents' generation. It will be their country soon. This gives me hope for the future.

Acknowledgements

This book depends heavily on the help that I received from numerous people all over the country who have shared their time and information with me or showed me their restoration work at first hand. For this I am very grateful to Bridget Barry, Simon Berrow, Sam Birch, Paul Brooks, Eugene Costello, David Duggan, Katharine Duff, Ciarán Fallon, John Feehan, Rory Finnegan, Ann Fitzpatrick, Padraic Fogarty, Michael Hickey, Colin Kelleher, Michael Keegan, Mary Kelly Quinn, Helen Lawless, Maria Long, Ferdia Marnell, Jim Martin, Kate McAney, Fiona McAuliffe, Mark McCorry, Alan McDonnell, Jack McGauley, Hugh McLindon, Derek McLoughlin, Rachel Millar, Ian Montgomery, Evelyn Moorkens, Liam Morrison, Tony Nagle, Derry Nairn, Tim Nairn, Brian Nelson, Stephen Newton, Micheal Ó Briain, Oliver Ó Cadhla, Michael O'Clery, William O'Connor, Jack O'Donovan Trá, Ciaran O'Keeffe, Denise O'Meara, Cathal and Bronagh O'Rourke, Aileen O'Sullivan, Brian O'Toole, Gareth Pedley, Francis Ratnieks, Neil Reid,

Julian Reynolds, Jenni Roche, Eimear Rooney, Marc Ruddock, Donal Sheehan, David Smyth, Andreea Spinu, Michael Stinson, Denis Strong, Gilly Taylor, Chris Uys, Graeme Warren, and Juan Yanez.

For their help in reviewing individual chapters, my special thanks are due to Maurice Eakin, Ciarán Fallon, Helen Lawless, Liam Lysaght, Micheal Ó Briain, Jack O'Donovan Trá, Ken Whelan and Paddy Woodworth. Their comments and corrections added substantially to the text. As with some of my previous books, I am ever grateful to Sheila Armstrong who has done a great job editing the text and made lots of constructive comments. Cassia Gaden Gilmartin read the draft text and gave it the 'thumbs up.' I thank Aoife K. Walsh and all at New Island Books for their faith in the value of this book.

I was accompanied on many of the site visits for this book by my wife Wendy who was infinitely patient with my frequent absorption in the research and writing.

References

1   Leopold, A. (1949). *A Sand County Almanac: And sketches from here and there.* New York and Oxford. Oxford University Press.

2   Mabey, R. (2010). *Weeds: How vagabond plants gatecrashed civilisation and changed the way we think about nature.* London. Profile Books.

3   Woodworth, P. (2023). The critical distinction between 'nature restoration' and 'rewilding'. *The Irish Times.* 17 August 2023.

4   Hayden, T. & Harrington, R. (2000). *Exploring Irish Mammals.* Dublin. Town House & Country House.

5   Thompson, W. (1850). *The Natural History of Ireland.* Vol II. Birds. London. Reeve, Benham & Reeve.

6   Molloy, J. (2006). *The Herring Fisheries of Ireland (1900–2005).* Galway. Marine Institute.

7   Mountfort, G. (1958). *Portrait of a Wilderness: The story of the Coto Donana expeditions.* London. Hutchinson.

8   Mitchell, F. (1976). *The Irish Landscape.* London. Collins.

9   Praeger, R.L. (1937). *The Way That I Went.* Dublin. Hodges Figgis.

10　National Parks & Wildlife Service (2018). *Wild Nephin Wilderness Area Conversion Plan Phase 1 (2018–2033) Executive Summary*. Dublin. Department of Culture, Heritage and the Gaeltacht.

11　Tiernan, D. (2008). *Redesigning afforested western peatlands in Ireland*. Castlebar. Coillte Teoranta.

12　European Commission (2022). *Guidelines on Wilderness in Natura 2000: Management of terrestrial wilderness and wild areas within the Natura 2000 Network*. Brussels. European Union.

13　Leopold, A. (1949). *Op. cit.*

14　Chapman, P.J. & Chapman, L.L. (1982). *Otter Survey of Ireland 1980–81*. London. Vincent Wildlife Trust.

15　Marren, P. (2002). *Nature Conservation: A review of the conservation of wildlife in Britain 1950–2001*. London. Harper Collins.

16　Viney, M. (2003). *Ireland: A Smithsonian Natural History*. Belfast. Blackstaff Press.

17　O'Gorman, F. and Wymes, E. eds. (1973). *The Future of Irish Wildlife: a blueprint for development*. Dublin. An Foras Talúntais.

18　Cabot, D. (1999). *Ireland*. New Naturalist Series. London. Harper Collins.

19　Monbiot, G. (2013). *Feral: Rewilding the land, sea and human life*. London. Penguin Books.

20　Lynn, D. & O'Neill, F. eds. (2019). *The Status of EU Protected Habitats and Species in Ireland*. Volume 1: Summary Overview. Dublin. National Parks and Wildlife Service.

21　Gilbert, G., Stanbury, A. & Lewis, L. (2021). Birds of Conservation Concern in Ireland 4: 2020–2026. *Irish Birds* 43: pp. 1–22.

22 Long, M. (2023). Our Gorgeous Grasslands. *Irish Wildlife Magazine.* Summer 2023: pp. 20–21. Dublin. Irish Wildlife Trust.

23 Woodworth, P. (2013). *Our Once and Future Planet: Restoring the world in the climate change century.* Chicago and London. University of Chicago Press.

24 Everett, N. (2015). *The Woods of Ireland: A history, 700–1800.* Dublin. Four Courts Press.

25 Cooney, G. (2000). *Landscapes of Neolithic Ireland.* London. Routledge.

26 McMahon, P. (2023). *Island of Woods: How Ireland lost its forests and how to get them back.* Dublin. New Island Books.

27 Cambrensis, G. (c.1200). *The History and Topography of Ireland.* Translated by J.J. O'Meara 1982. Portlaoise. Dolmen Press.

28 McCracken, E. (1971). *Irish Woods since Tudor Times: Distribution and exploitation.* Newton Abbot. David and Charles.

29 Perrin, P.M. & Daly, O.H. (2010). *A provisional inventory of ancient and long-established woodland in Ireland. Irish Wildlife Manuals,* No. 46. Dublin. National Parks and Wildlife Service.

30 Perrin, P., Martin, J., Barron, S., O'Neill, F., McNutt, K. & Delaney, A. (2008). *National Survey of Native Woodlands 2003–2008.* Dublin. National Parks and Wildlife Service.

31 Roche, J.R., Mitchell, F.J.G., Waldren, S., Stefanini, B.S. (2018). Palaeoecological evidence for survival of Scots Pine through the late Holocene in Western Ireland: implications for ecological management. *Forests* 9: p. 350.

32 Belton, S., Cubry, P., Roche, J.R. and Kelleher, C.T. (2024). Molecular characterisation of *Pinus sylvestris* (L.) in Ireland

at the western limit of the species distribution. *BMC Ecology and Evolution* 24: p.12.

33   Daltun, E. (2022). *An Irish Atlantic Rainforest: A personal journey into the magic of rewilding*. Dublin. Hachette Books.

34   Ashmole, P. & Ashmole, M. (2020). *A Journey in Landscape Restoration: Carrifran Wildwood and beyond*. Caithness. Whittles Publishing.

35   Carey, M. (2009). *If Trees Could Talk: Wicklow's trees and woodland over four centuries*. Dublin. Coford.

36   Scott, D. (2022). Ancient jewels in a Northern Irish landscape. In: *Wood Wise: Nature recovery at scale* (ed. K. Hornigold). Grantham. The Woodland Trust.

37   Hayden, T. & Harrington, R. (2000). *Op. cit.*

38   Purser, P., Wilson, F. and Carden, R. (2009). *Deer and forestry in Ireland: A review of current status and management requirements*. Report to Woodlands of Ireland.

39   Irish Deer Management Strategy Group (2023). *Final Report into Developing a Sustainable Deer Management Strategy for Ireland*. Dublin. Department of Agriculture, Food & the Marine.

40   Morera-Pujol, V., Mostert, P.S., Murphy, K.J., Burkitt, T., Coad, B., McMahon, B.J., Nieuwenhuis, M., Morelle, K., Ward, A.I. & Ciuti, S. (2022). Bayesian species distribution models integrate presence-only and presence–absence data to predict deer distribution and relative abundance. *Ecography* 2023, pp.1–14.

41   Tees, A. & Tees, M. eds. (2022). *Irish Peaks: A celebration of Ireland's highest mountains*. Dublin. Mountaineering Ireland.

42   Perrin, P.M., Barron, S.J., Roche, J.R. & O'Hanrahan, B. (2014). *Guidelines for a national survey and conservation*

assessment of upland vegetation and habitats in Ireland.
Version 2.0. Irish Wildlife Manuals, No. 79. Dublin.
National Parks and Wildlife Service.

43   Costello, E. (2020). Hill farmers, habitats and time: the
     potential of historical ecology in upland management and
     conservation. *Landscape Research* 45, pp. 951–965.

44   Colhoun, K. (2023). Red Grouse. *Biodiversity Ireland* 24.
     Spring Summer 2023. p.5

45   Fogarty, P. (2017). *Whittled Away: Ireland's Vanishing
     Nature.* Cork. Collins Press.

46   McAuliffe, F., Wilson-Parr, R., O'Toole, L., Marnell, F. &
     Reid, N. (in press). Prey deficit for reintroduced Golden
     Eagles (*Aquila chrysaetos*) in Ireland. *Irish Naturalists'
     Journal.*

47   O'Neill, F.H., Martin, J.R., Devaney, F.M. & Perrin, P.M.
     (2013). *The Irish semi-natural grasslands survey 2007–2012.*
     Irish Wildlife Manuals, No. 78. Dublin. National Parks and
     Wildlife Service.

48   Colhoun, K., Mawhinney, K., McLaughlin, M., Barnett,
     C., McDevitt, A., Bradbury, R.B. & Peach, W. (2017). Agri-
     environment scheme enhances breeding populations of some
     priority farmland birds in Northern Ireland, *Bird Study,* 64:
     pp. 545–556.

49   Harrison, S., Kelly, S.B.A., & O'Donoghue, B.G., (2023).
     Curlew Conservation Programme Annual Report 2023.
     Dublin. National Parks and Wildlife Service.

50   Roche, N., Aughney, T., Marnell, F. & Lundy, M. (2014).
     Irish Bats in the 21st Century. Virginia. Bat Conservation
     Ireland.

51   Lewis, L.J., Coombes, D., Burke, B., O'Halloran, J., Walsh,
     A., Tierney, T. D. & Cummins, S. (2019). Countryside

Bird Survey: Status and trends of common and widespread breeding birds 1998–2016. *Irish Wildlife Manuals,* No. 115. Dublin. National Parks and Wildlife Service.

52 Reid, N. (2018). Running with the Hare. *Biodiversity Ireland* 18: pp. 14–15. Waterford. National Biodiversity Data Centre.

53 Evans, E.E. (1957). *Irish Folk Ways.* London. Routledge & Kegan Paul.

54 Kelly Quinn, M., Feeley, H. and Bradley, C. (2020). Status of freshwater invertebrate biodiversity in Ireland's rivers – time to take stock. *Biology and Environment: Proceedings of the Royal Irish Academy* 120B: pp. 65–82.

55 Kelly Quinn *et al.* (2020). *Ibid.*

56 Osbourne, L.L. and Kovacic, D.A. (1993). 'Riparian vegetated buffer strips in water-quality restoration and stream management'. *Freshwater Biology* 29: pp. 243–258.

57 Loerke, E., Ina Pohle, I., Wilkinson, M.E., Rivington, M., Wardell-Johnson, D. & Geris, J. (2023). Long-term daily stream temperature record for Scotland reveals spatio-temporal patterns in warming of rivers in the past and further warming in the future. *Science of The Total Environment* 890: pp. 164-194.

58 O'Grady, M.F. (2006). *Channels and Challenges: Enhancing Salmonid Rivers.* Irish Freshwater Ecology & Management Series. Number 4. Dublin. Central Fisheries Board.

59 Feehan, J. & O'Donovan, G. (1996). *The Bogs of Ireland: An introduction to the natural, cultural and industrial heritage of Irish peatlands.* Dublin. University College Dublin.

60 Cambrensis, G. (c.1200). *Op. cit.*

61 D'Arcy, G. (1999). *Ireland's Lost Birds.* Dublin. Four Courts Press.

62 Whelan, R. (2017). Community conservation at Abbeyleix Bog. *Irish Wildlife*. Autumn 2017. Irish Wildlife Trust.

63 Feehan, J. (2021). *When the Nightjar Returns ...: The natural history and human story of Killaun Bog, County Offaly*. Birr. St. Brendan's Community School and Offaly County Council.

64 Clover, C. (2022). *Rewilding the Sea: How to save our oceans*. London. Penguin Random House.

65 Bolster, W.J. (2014). *The Mortal Sea: Fishing the Atlantic in the age of sail*. Cambridge, Massachusetts. Harvard University Press.

66 Fairley, J. (1981). *Irish Whales and Whaling*. Belfast. Blackstaff Press.

67 Blázquez Hervás, M., Whooley, P. , Massett, N., O'Brien, J., Wenzel, F., O'Connor, I. & Berrow, S. (2023) Abundance estimates of humpback whales *(Megaptera novaeangliae)* in Irish coastal waters using mark recapture and citizen science. *Journal of Cetacean Research and Management* 24: pp. 209–225.

68 Berrow, S. & Whooley, P. (2022). Managing a dynamic North Sea in the light of its ecological dynamics: increasing occurrence of large baleen whales in the southern North Sea. *Journal of Sea Research* 182: pp. 102–186.

69 Cummins, S., Lauder, C., Lauder, A. & Tierney, T.D.(2019). *The Status of Ireland's Breeding Seabirds: Birds Directive Article 12 Reporting 2013–2018*. Irish Wildlife Manuals, No. 114. Dublin. National Parks and Wildlife Service.

70 Cummins, S., Lewis, L.J. & Egan, S. (2016). *Life on the Edge: Seabird and fisheries in Irish waters*. Kilcoole. BirdWatch Ireland.

71 Collins,T., Malone, J. & White, P. (2006). *Report of the Independent Salmon Group*. A Report to Minister for State

at the Department of Communications, Marine and Natural Resources, John Browne T.D.

72   Siggins, L. (2017). Driftnet ban fails to save Atlantic wild salmon. *The Irish Times*. 19 February 2017.

73   Wilkins, N.P. (2004). *Alive Alive O: The shellfish and shellfisheries of Ireland*. Galway. Tir Eolas.

74   Hession, C., Guiry, M.D., McGarvey, S. & Joyce, D. (1998). *Mapping and assessment of the seaweed resources off the west coast of Ireland. Marine Resource Series No. 5*. Dublin. Marine Institute.

75   Clover, C. (2022) *Op. cit.*

76   Eger, A.M., Layton, C., McHugh, T.A., Gleason, M., & Eddy, N. (2022). *Kelp Restoration Guidebook: Lessons Learned from Kelp Projects Around the World*. The Nature Conservancy, Arlington, USA.

77   McNally, K. (1976). *The Sun-fish Hunt*. Blackstaff Press. Belfast.

78   Marine Protected Area Advisory Group (2020). *Expanding Ireland's Marine Protected Area Network: A report by the Marine Protected Area Advisory Group*. Report for the Department of Housing, Local Government and Heritage, Ireland.

79   Stewart, B.D., Howarth, L.M., Wood, H., Whiteside, K., Carney, W., Crimmins, É., O'Leary, B.C., Hawkins, J.P. and Roberts, C.M. (2020). Marine Conservation Begins at Home: How a local community and protection of a small bay sent waves of change around the UK and beyond. *Frontiers in Marine Science*. 7:76. doi: 10.3389/fmars.2020.00076.

80   Clover, C. (2022) *Op. cit.*

81   Warren, G. (2022). *Hunter-gatherer Ireland: Making connections in an island world*. Oxford. Oxbow Books.

82   Montgomery, W.I. (2014). Origin of the Holocene mammals of 'these islands'. In: *Mind the Gap II: new insights into the Irish postglacial* (edited by D.P. Sleeman, J. Carlsson & J.E.L. Carlsson). Belfast. Irish Naturalists' Journal.

83   Warren, G. (2022). *Op. cit.*

84   Woodman, P., McCarthy, M. & Monaghan, N. (1997). The Irish Quaternary Fauna Project. *Quaternary Science Reviews* 16: pp. 129–159.

85   Hickey, K. (2011). *Wolves in Ireland: A natural and cultural history*. Dublin. Four Courts Press.

86   D'Arcy, G. (1999). *Op. cit.*

87   National Biodiversity Data Centre. 10 species Ireland has lost. https://biodiversityireland.ie/top10/10-species-ireland-has-lost.

88   National Biodiversity Data Centre. *Ibid.*

89   Lucey, J. (2005). *The Irish Pearl: A cultural, social and economic history*. Bray. Wordwell.

90   Faulkner, J. (2023). *Ireland's Changing Flora: A Summary of the Results of Plant Atlas 2020*. Durham. Botanical Society of Britain and Ireland.

91   Ussher, R.J. & Warren, R. (1900). *Birds of Ireland*. London. Gurney and Jackson.

92   Rooney, E. (2013). *Ecology and Breeding Biology of the Common Buzzard* Buteo buteo *in Ireland*. PhD thesis. Belfast. Queens University Belfast.

93   Balmer, D.E., Gillings, S., Caffrey, B.J., Swann, R.L., Downie, I.S. & Fuller, R.J. (2013). *Bird Atlas 2027–11: the breeding and wintering birds of Britain and Ireland*. Thetford. BTO Books.

94   Ó Crohán, T. (1934). *The Islandman*. Dublin. Talbot Press

95   Hutchinson, C.D. (1989). *Birds in Ireland*. Calton. Poyser.

96   Newton, S.F., Harris, M.P. & Murray, S. (2015) Census of Gannet *Morus bassanus* colonies in Ireland in 2013–2014. *Irish Birds* 10: pp. 215–220.

97   Wanless & Harris (2004). Northern Gannet in: *Seabird Populations of Britain and Ireland*. (eds. Mitchell, P.I., Newton, S.F., Ratcliffe, N. & Dunn, T.E.). Calton. Poyser.

98   Morris, C.D. & Duck, C.D. (2019). *Aerial thermal-imaging survey of seals in Ireland, 2017 to 2018. Irish Wildlife Manuals*, No. 111. Dublin. National Parks and Wildlife Service.

99   Natural Environment Research Council (2021). *Scientific Advice on Matters Related to the Management of Seal Populations:* University of St. Andrews. Special Committee on Seals. NERC Sea Mammal Research Unit.

100   Scharff, R.F., Ussher, R.J., Cole, G.A.J., Newton, E.T., Dixon, A.F and Westropp, T.J. (1906). The exploration of the caves of County Clare. *Transactions of the Royal Irish Academy* 33B: 1–76

101   McDevitt, A.D., Kajtoch, L., Mazgajski, T.D., Carden R.F., Coscia I., Osthoff, C., Coombes, R.H. & Wilson, F. (2011). The origins of Great Spotted Woodpeckers *Dendrocopos major* colonizing Ireland revealed by mitochondrial DNA. *Bird Study* 58: pp. 361– 364

102   Ratnieks, F.L.W., Beckett, O., Nelson, B. and FitzPatrick, Ú. (2022). Distribution and status of the Ivy Bee *(Colletes hederae)* in Counties Wexford and Wicklow, Ireland, Autumn 2022. *Irish Naturalists' Journal* 39: pp. 67–75

103   Faulkner, J. (2023). *Op. cit.*

104   D'Arcy, G. (1999). *Op. cit.*

105   IUCN/SSC (2013). *Guidelines for Reintroductions and Other Conservation Translocations*. Version 1.0. Gland, Switzerland: IUCN Species Survival Commission.

106 Mee, A., Breen, D., Clarke, D., Heardman, C., Lyden, J., McMahon, F., O'Sullivan, P. & O'Toole, L. (2016). Reintroduction of White-tailed Eagles *Haliaeetus albicilla* to Ireland. *Irish Birds* 10: pp. 301–314.

107 Mee, A., Heardman, C., Clarke, D., Nygård, T., O'Toole, L. & Buckley, P. (2023). White-tailed Sea Eagle *Haliaeetus albicilla* reintroduction to Ireland: restoration of a large avian apex predator. Abstract of paper presented at CORC 2023. 8th Irish Ornithological Research Conference. University College Cork. *Irish Birds* 45: pp. 134-135.

108 Murray, T. & Wyse Jackson, M. (2022). The history, status and conservation management of Cottonweed *Achillea maritima (Otanthus maritimus)* (Asteraceae) at Lady's Island Lake, Co. Wexford, Ireland *British & Irish Botany* 4: pp. 248–272.

109 Praeger, R.L. (1937). *Op. cit.*

110 Cambrensis, G. (c.1200). *Op. cit.*

111 Kelly, F., King, J., Gargan, P. & Roche, W. (2020). Fish in Irish Rivers. In: *Ireland's Rivers* (ed: M. Kelly-Quinn and J. Reynolds). Dublin. University College Dublin Press.

112 Beebee, T.J.C. (2002). The Natterjack toad *(Bufo calamita)* in Ireland: Current status and conservation requirements. *Irish Wildlife Manuals,* No. 10. Dublin. Dúchas the Heritage Service.

113 King, J.L., Marnell, F., Kingston, N., Rosell, R., Boylan, P., Caffrey, J.M., FitzPatrick, Ú., Gargan, P.G., Kelly, F.L., O'Grady, M.F., Poole, R., Roche, W.K. & Cassidy, D. (2011). *Ireland Red List No. 5: Amphibians, Reptiles & Freshwater Fish.* Dublin. National Parks and Wildlife Service.

114 Jepson, P. & Blythe, C. (2020). *Rewilding: The radical science of ecological recovery.* London. Icon Books.

115 Twi nings, J.P., Montgomery, W.I. & Tosh, D.G. (2021). Declining invasive grey squirrel populations may persist

in refugia as native predator recovery reverses squirrel species replacement. *Journal of Applied Ecology* 58: pp. 248–260.

116 Bavin, D. & MacPherson, J. (2022). *The Lynx to Scotland Project: Assessing the social feasibility of potential Eurasian lynx reintroduction to Scotland.* Vincent Wildlife Trust.

117 Lysaght, L. & Marnell, F. (2016). *Atlas of Mammals in Ireland 2010–2015.* Waterford. National Biodiversity Centre.

118 McDevitt, A., Carden, R., Coscia, I. & Frantz, A. (2013). Are wild boars roaming Ireland once more? *European Journal of Wildlife Research* 59: pp. 761–764.

119 Dennis, R. (2021). *Restoring the Wild: Rewilding our skies, woods and waterways.* London. William Collins.

120 D'Arcy, G. (1999). *Op. cit.*

121 Anon. (2023). Ospreys breeding in Ireland after 200 years 'like finding long-lost treasure'. *The Irish Times*, 24 August 2023.

122 Law, A., McLean, F. & Willby, N.J. (2016). Habitat engineering by beaver benefits aquatic biodiversity and ecosystem processes in agricultural streams. *Freshwater Biology* 61: pp. 486–499.

123 Monbiot, G. (2013). *Op. cit.*

124 Viney, M. (2003). *Op. cit.*

125 Citizens' Assembly (2023). *Report of the Citizens' Assembly on Biodiversity Loss.* Dublin. Citizens' Assembly.

126 Featherstone, A.W. (2023). Restoring the Ancient Caledonian Forest. TEDx talk. https://alanwatsonfeatherstone.com/speaking/tedx-talk/

127 Woodworth, P. (2013). *Op. cit.*

128 Tree, I. (2018). *Wilding: The return of nature to a British farm.* London. Picador.

Index